CONTROLLING WITH COMPUTERS
Control theory and practical digital systems

John Billingsley

Professor of Robotics
Portsmouth Polytechnic

McGRAW-HILL BOOK COMPANY

London · New York · St Louis · San Francisco · Auckland
Bogotá · Guatemala · Hamburg · Lisbon · Madrid · Mexico
Montreal · New Delhi · Panama · Paris · San Juan · São Paulo
Singapore · Sydney · Tokyo · Toronto

Published by
McGRAW-HILL Book Company (UK) Limited
MAIDENHEAD · BERKSHIRE · ENGLAND

British Library Cataloguing in Publication Data
Billingsley, John
　　Controlling with computers.
　　1. Control theory. Applications of, in
　　computer systems
　　I. Title
　　629′.8′312
　　ISBN 0-07-084193-4

Library of Congress Cataloging-in-Publication Data
Billingsley, J. (John)
　　Controlling with computers: control theory and practical digital
　　systems / John Billingsley.
　　　　p.　　　　cm.
　　Includes index.
　　ISBN 0-07-084193-4
　　1. Automatic control.　2. Control theory.　3. Digital control
　　systems.　I. Title.
　　TJ223.M53B54 1989
　　629.8′95—dc19　　88-13410

1234 CUP 9089

Typeset by Eta Services (Typesetters) Ltd, Beccles, Suffolk
Printed and bound in Great Britain at the University Press, Cambridge

CONTENTS

v

PREFACE

I am always suspicious of a textbook that promises that a subject can be 'made easy'. Control theory is not an easy subject, but it is a fascinating one. It embraces every phenomenon that is described by its variation with time, from the trajectory of a projectile to the vagaries of the stock exchange. Its principles are as essential to the ecologist and the historian as they are to the engineer.

All too many students regard control theory as a rag-bag of party tricks for production in examinations. 'Learn how to plot a root locus, and question three is easy.' Frequency domain and time domain methods are pitted against each other as alternatives, and somehow the spirit of the subject falls between the cracks.

Control theory is a story with a plot. State equations and transfer functions all lead back to the same point: to an encapsulation of the actual system that they have been put together to describe.

The subject is certainly not one that can be made easy, but perhaps the early, milder chapters will give the novice an appetite for the tougher meat which follows. The intention of the book is to explain and convince rather than to drown the reader in detail, and the mathematical introductions are designed to smooth the path for each successive chapter.

The earlier material has been 'test driven' on final-year undergraduates, and I am grateful to the members of the M.Sc. course who proof-read the whole of the text (or so they have assured me). I would like to think that the progressive nature of the mathematics could open up the early material to students of school physics and computing—but maybe that is hoping too much.

The computer certainly plays a large part in appreciating the material. With the aid of a few lines of software and a trusty PC, the reader can simulate dynamic systems in real time. Other programs, small enough to type into the machine in a few minutes, give access to on-screen graphical analysis methods including Bode, Nyquist, Nichols and Root Locus in both s and z planes. Indeed, using the "GRAPHICS" command to obtain a screen dump, many of the illustrations were produced from the very programs that are listed here.

There are many people to whom I owe thanks for this book. First I must mention Professor John Coales, who guided my research in Cambridge. I am indebted to many colleagues over the years, both in industry and academe. I probably learned most from those with whom I disagreed most strongly!

I cannot lose the opportunity to thank two schoolteachers, Eric Barton and Freddy Nutbourne, who opened up the world of mathematics.

My wife, Rosalind, has kept up a supply of late-night coffee and encouragement while I have pounded the text into an antique Commodore which still serves faithfully as a word-processor. The illustrations are all drawn and a host of errors have been corrected. Now it is all up to the publishers—and to the readers!

John Billingsley

CONTROL THEORY, ART AND PRACTICE

1-1 INTRODUCTION

The most straightforward approach to a subject is rarely in the order of its historical development. Few electronics courses start off by rubbing a piece of amber, or even consider a coherer. The young mathematician is unlikely ever to meet a quaternion, however familiar he (or she) may become with complex numbers. Control theory is no exception to the rule.

In the early days, control theory amounted to simple feedback analysis. Mix some of the output signal back into the input, and some of the inconsistencies of an amplifier could be ironed out. Increase the amount of feedback, and the result improved up to a point; then the amplifier would start to ring or oscillate. The same was true of a position controller, say for automatic aiming of a gun. As the feedback was increased, so was the accuracy and speed of response, until once again instability gave trouble.

Control theory concerned itself with determining the maximum 'feedback gain' which could be applied before instability started to arise. To perform an analysis, some means had to be found to provoke the system into revealing its nature. A step of input signal might be applied, but how could the response be measured? Oscilloscopes did not at first exist, and if the early pen recorders did not smother you in ink their response owed as much to their own dynamics as to the system under test. The easiest way to test a system was instead to whistle at it.

A sine wave 'whistle' from an oscillator could be injected into the system, using a simple a.c. voltmeter to measure the resulting output. If the same voltmeter was used to measure input and output at an assortment of frequencies, any errors arising from variation of voltmeter sensitivity with frequency would be cancelled out. The 'frequency response' became the yardstick of system performance. A variety of techniques rapidly followed for predicting the 'closed loop' response from the 'open loop' frequency response. With the oscilloscope came the possibility of measuring phase shift in addition to response amplitude, and yet more graphical techniques were developed.

Newer instruments came onto the market, such as the 'wobbulator', popular for setting up television receivers, which injected a succession of rising whoops of sine waves and displayed the entire frequency response as a trace on an oscilloscope. The 'transfer function analyser' was used for testing servomotor responses; it injected a sine wave input signal and at the same time broke the received output signal into terms from which both amplitude and phase shift could be deduced. All these techniques were firmly rooted in the 'frequency domain'.

Aircraft demanded a more sophisticated breed of controller. Even the simple 'George' autopilot needed some dynamics in its control. One aircraft instrument, developed around the mid-century, was known as 'Windy Willie'. It was powered by venturi vacuum, and had an ingenious assortment of valves and pistons to amplifv and integrate signals. For precise computation of aircraft track for bomb aiming, another form of integrator was used, the 'ball and plate'.

With integrators available as mechanical units and then as electronic circuits, the behaviour of a system could start to be simulated by setting up an array of connected integrators that satisfied the same equations. The analogue computer was relied on more and more to try out control designs before flying them. To make the simulation easy, the system equations had to be rearranged into a set of simple expressions representing the mixture of signals to be applied to each integrator.

With the advent of the low-cost digital computer, simulation by digital means became more attractive than analogue. Meanwhile the representation of the system as a collection of first-order equations was seen in a new light. The integrator outputs could be represented as a mathematical vector, the equations fell into the shape of a 'matrix state equation' and a whole new range of techniques emerged.

At first these methods were regarded as advanced and difficult, to be covered late in a course on control theory. It became clear, however, that because of the ease of representing simulations on a desk-top computer, complete with graphic output, this approach gave an easily approachable insight and understanding. This book therefore starts by breaking systems into collections of simple first-order equations, so that elementary computer programs can simulate their behaviour.

Once the scene has been set, it is possible to represent the equations in a formal, matrix way, and to apply algebra to the task of analysing the effect of feedback. Now the system takes on a graspable identity, against which the sinusoidal test methods can be set, rather than remaining a mysterious 'black box'. Time and frequency domain methods are examined side by side, first in continuous time and then in the discrete time world of computer control. On the way, it is recognized that very many control systems apply non-linear control, either through drive saturation or with the deliberate aim of improving performance. The final chapter takes a look at optimal control, both linear and 'bang-bang'.

As the subject progresses, a growing battery of mathematical tools will be needed. It is always irksome to have to turn to an appendix for clarification of a mathematical point; an appendix somehow gives the impression that it is not meant to be read at all. Necessary items of theory are therefore introduced here in 'preludes' before most chapters. It does no harm to scan through them, however familiar the material may appear, and they can be easily skipped on a second reading.

It can be irritating to have the argument of a book interrupted by a set of aggressive questions each time the author makes a point. On the other hand, publishers insist on them, so examples appear through the book. Their solutions will often be found in the text that follows them, although some problems are left unsolved until a later chapter.

The early chapters should be fairly easy for undergraduates or for engineers whose mathematics have become rusty. However, the final target is an advanced one. By no means is the book a mere light introduction; it is intended to lead the reader to a sound understanding of control principles and techniques, which is at the same time both broad and deep.

DIFFERENTIAL EQUATIONS

P2-1 INTRODUCTION

The primary concern of control theory will be seen to be the behaviour of a system with time. We must therefore establish some notation and conventions for dealing with concepts such as *rate of change*.

If the value of some variable is represented by x and if x is a function of time, then we might write it as $x(t)$, or simply as x, as the occasion arises. A moment later, the time will have moved on from t to $t + \delta t$, by which time x will have changed its value to $x(t + \delta t)$. The change over the interval is given by

$$\delta x = x(t + \delta t) - x(t)$$

The rate of change of x is indicated by the ratio of this change in x to the change δt in time, examined as the time interval is made smaller and smaller. The rate of change, or derivative with respect to time, is defined by the limit as δt tends to zero:

$$\frac{\mathrm{d}x}{\mathrm{d}t} = \lim_{\delta t \to 0} \frac{x(t + \delta t) - x(t)}{\delta t}$$

It will often be convenient to write this rate of change in a shorthand form, simply by putting a 'dot' over the variable, i.e.

$$\dot{x} \text{ is the same as } \frac{\mathrm{d}x}{\mathrm{d}t}$$

5

To be more precise, we should be concerned with $\dot{x}(t)$, and it is at once obvious that we can consider the rate of change of this new variable and its rate-of-change-of-rate-of-change to any degree possible. Two dots give the second derivative, three the third and so on.

Differentiation can also be represented as the application of the 'differential operator' D. A signal and its derivatives will now appear as x, Dx, D^2x, D^3x and so on. What starts out as convenient notation soon leads to some powerful techniques. It is easy to show that combinations of constants and Ds can be multiplied out; for example:

$$(D + 2)(D + 3)x = D^2x + 5Dx + 6x$$

$$= \ddot{x} + 5\dot{x} + 6x$$

Having succeeded in multiplying such expressions, it is tempting to try dividing them; this is where mistakes can occur. If we have an equation:

$$(D + 3)x = y$$

then we might reasonably write

$$x = \frac{1}{D + 3}y$$

as long as we strictly restrict our interpretation to mean that x is one of the many solutions to

$$\dot{x} + 3x = y$$

The solution includes a 'complementary function', here taking the value of $A\,e^{-3t}$, where A is a constant adjusted to fit the initial value of x. Above all we must resist the temptation to infer from

$$(D + 3)x = (D + 3)y$$

that x must be equal to y; this is only one of the many possible solutions.

To put the algebraic treatment of differential equations onto a firmer footing, Laplace transform notation can be used. The Laplace transform can be a happy mathematicians' playground of infinite integrals of functions of a complex variable. For the control engineer, however, it is really no more than a disciplined way of unscrambling operators while providing a neat method of dealing with initial conditions.

MODELLING TIME

2-1 INTRODUCTION

In every control problem, time is involved in some way. It might appear in an obvious way relating the height at each instant of a spacecraft, in a more subtle way as a list of readings taken once per week or unexpectedly as a feedback amplifier bursts into oscillation.

Occasionally time may be involved as an explicit function, such as the height of the tide at four o'clock, but more often its involvement is through differential or difference equations, linking the system behaviour from one moment to the next. This is best seen with an example.

2-2 A SIMPLE SYSTEM

A cup of coffee has just been made. It is rather too hot at the moment, at 80°C. If left for some hours it would cool down to room temperature at 20°C, but just how fast would it cool and when would it be at 60°C?

It is a reasonable assumption that the rate of loss of heat is proportional to the temperature above ambient. The rate of loss of heat will in turn be proportional to the rate of fall in temperature, and so we see that

$$\frac{dT}{dt} = -k\,(T - T_{\text{ambient}})$$

If we can determine the value of the constant k, perhaps by a simple experiment, then the equation can be solved for any particular initial temperature—the form of the solution comes later.

Equations of this sort apply to a vast range of situations. A rainwater butt has a small leak at the bottom (Fig. 2-1). The rate of leakage is proportional to the depth, H, and so

$$\frac{dH}{dt} = -kH$$

The water will leak out until eventually the butt is empty. But suppose now that there is a steady flow *into* the butt, sufficient to raise the level (without leak) at a speed u. Then the equation now becomes

$$\frac{dH}{dt} = -kH + u$$

Figure 2-1 Rainwater butt with small leak.

What will the level of the water settle down at now? When it has reached a steady level, however long it takes, the rate of change of depth will have fallen to zero, so $dH/dt = 0$. It is not hard to see that $-kH + u$ must also be zero, and so $H = u/k$.

Now if we really want to know the depth as a function of time, a mathematical formula can be found for the solution. But let us try another approach first—simulation.

2-3 SIMULATION

With very little effort, we can construct a computer program that will imitate the behaviour of the bucket. If the depth right now is H, then we have already described the rate of change of depth dH/dt as $(-kH + u)$. In a short time dt, the depth will have changed by

$$(-kH + u)\, dt$$

so that in program terms we have

H = H + (−k∗H + u)∗dt

This will work as it stands as a line of BASIC or FORTRAN, although some machines might insist on upper case letters throughout; for Pascal or ALGOL add a colon and a semicolon as needed.

Even when wrapped up in input and output statements to make a complete program, the simulation is very simple. For almost any micro, the following program will give a primitive screen display:

```
 10 INPUT "Initial level 0 to 40"; H
 20 INPUT "Input u, 0 to 20    "; U
 30 K  = 0.5
 40 DT = 0.01
 70 REM This is the simulation loop
 80 H = H + (−K∗H + U)∗DT
 90 PRINT TAB(H); "∗"
100 GOTO 70
```

As an exercise, you might modify line 90 and add another line or two to give a more usable graphical plot of the result. Your program will probably include $T = T + DT$ and the equivalent of DRAW(T∗TSCALE, H∗HSCALE).

When you run such a simulation program, you will notice that the result will vary if you change the value of DT—even when H is plotted against time, rather than step number. For small step lengths the change might be negligible, but for values approaching one second the approximation deteriorates rapidly; try DT = 4 and see what happens!

Shortly we will see how to perform an accurate computer simulation

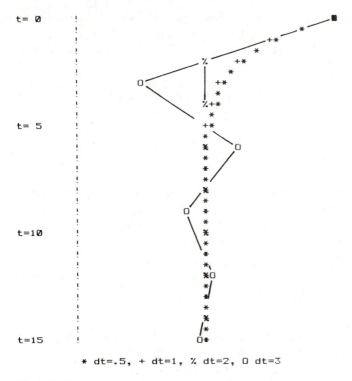

* dt=.5, + dt=1, % dt=2, O dt=3

Figure 2-2 Elementary simulation—the effect of changing step length.

with no limit on step size, provided that the input does not change between steps. Examples of simulation are given in Fig. 2-2.

2-4 SOLVING THE FIRST-ORDER EQUATION

At last we must consider the formal solution of the simple example. The treatment here may seem overelaborate, but later on we will apply the same methods to more demanding systems.

By using the variable x instead of H or T_{coffee}, we can put both simple examples into the same form:

$$\frac{\mathrm{d}x}{\mathrm{d}t} = ax + bu \tag{2-1}$$

where a and b are constants that describe the system and u is an input, which can simply be a constant as T_{ambient} is in the first example.

Rearranging, we see that

$$\frac{dx}{dt} - ax = bu \tag{2-2}$$

Since we have a mixture of x and dx/dt, we cannot simply integrate. We must somehow find a function of x and time which will accommodate both terms on the left of the equation.

Let us consider

$$\frac{d}{dt}[xf(t)]$$

where $f(t)$ is some function of time which has derivative $f'(t)$. When we differentiate by parts we see that

$$\frac{d}{dt}[xf(t)] = \frac{dx}{dt}f(t) + xf'(t) \tag{2-3}$$

If we multiply Eq. (2-2) through by $f(t)$ in the hope of matching its left-hand side to the right-hand side of Eq. (2-3), then we find $x[-af(t)]$ where we would like to find $xf'(t)$. Clearly we need to choose $f(t)$ so that

$$f'(t) = -af(t)$$

i.e.

$$f(t) = e^{-at}$$

Now we get, from (2-2),

$$e^{-at}\frac{dx}{dt} - a\,e^{-at}x = e^{-at}bu \tag{2-4}$$

so we can write

$$\frac{d}{dt}(e^{-at}x) = e^{-at}bu \tag{2-5}$$

At last we can integrate both sides to get

$$[e^{-at}x]_0^t = \int_0^t e^{-at}bu\,dt \tag{2-6}$$

$$e^{-at}x(t) - 1 \cdot x(0) = \int_0^t e^{-at}bu\,dt \tag{2-7}$$

and so the solution is

$$x(t) = e^{at}x(0) + e^{at}\int_0^t e^{-at}bu\,dt \tag{2-8}$$

Now if u remains constant throughout the interval 0 to t, we can simplify this still further to

$$x(t) = e^{at} x(0) + hu \qquad (2\text{-}9)$$

where

$$h = e^{at} \int_0^t e^{-at} b \, dt$$

$$= (e^{at} - 1)\frac{b}{a} \qquad (2\text{-}10)$$

This then is our answer to the varying step length of the last section. If we compute a constant g from

$$g = e^{at}$$

together with h from Eq. (2-10), then the value of x is taken precisely from step to step by the relationship

$$x_{n+1} = gx_n + hu_n \qquad (2\text{-}11)$$

The simulation program of the last section can now be made more precise. The value of a is $-k$, so that for step length dt we have:

```
10 INPUT "Initial level 0 to 40"; X
20 INPUT "Input u, 0 to 20    "; U
30 K = 0.5
40 DT = 0.01
50 G = EXP(−K∗DT)
60 H = (EXP(−K∗DT) − 1)∗1/(−K)
70 REM This is the loop again
80 X = G∗X + H∗U
90 PRINT TAB(X); "∗"
100 GOTO 70
```

The magic is all in line 80. It almost looks too easy. However, can the method cope with higher order systems?

2-5 A SECOND-ORDER PROBLEM

A servomotor drives a robot axis to position x. The speed of the axis is v. The acceleration is proportional to the drive current u; for the moment there is no damping (see Fig. 2-3). Can we model the system to deduce its performance?

In Sec. 2-3 we had an equation for the rate of change of a variable which described the system. By repeatedly adding the small change over a short interval, we were able to track the progress of the system against time. Let us try the same approach here.

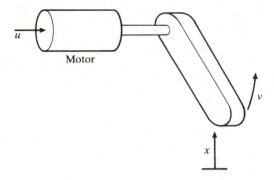

Figure 2-3 Position control example.

The system is described by x, so we look for an equation for dx/dt. When we spot it, it is simply

$$\frac{dx}{dt} = v$$

So what do we know about v? It would help to know its rate of change, which we have been told is proportional to u, the input:

$$\frac{dv}{dt} = bu$$

We therefore arrive at a pair of differential equations. If we know the initial values of x and v, we can recalculate the new values after each interval dt as before. The simulation program can now be written:

```
 10 INPUT "Initial position 0 to 40"; X
 15 INPUT "Initial speed >0        "; V
 20 INPUT "Input u, 0 to 20        "; U
 30 B = 0.1
 40 DT = 0.1
 50 REM Loop begins
 70 X = X + V*DT
 80 V = V + B*U*DT
 90 PRINT TAB(X); "*"
100 GOTO 50
```

The result of running it will not be very interesting. The drive u will cause the robot to accelerate, and the trail of asterisks will soon run off the screen. We really need to consider the performance under control, when the input u has by feedback been made proportional to the error between position x and some target xt:

$$U = F*(XT - X)$$

We will find that the system oscillates for ever unless some damping is added. A proportion of the speed v is added, to give:

$U = F*(XT - X) - D*V$

Once again the program is changed, to become:

```
10 INPUT "Initial position 0 to 40"; X
15 INPUT "Initial speed >0        "; V
20 INPUT 'Feedback, damping    "; F,D
25 INPUT "Target 0 to 30        "; XT
30 B = 0.1
40 DT = 0.1
50 REM Loop here
60 U = F*(XT - X) - D*V
70 X = X + V*DT
80 V = V + B*U*DT
90 PRINT TAB(X); "*"
100 GOTO 50
```

With a target of 30, try values for F and D of 60 and 20 (see Fig. 2-4). You can see the result of omitting damping by entering zero for D on another run. You might also like to tidy up the programs of this chapter, improving the plotting

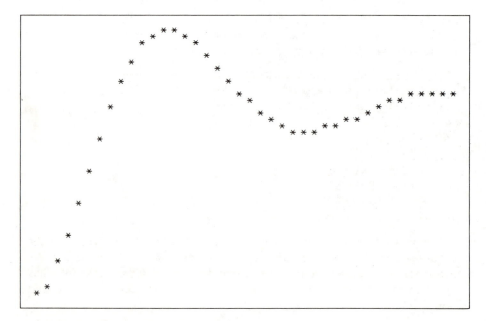

Figure 2-4 Simulation example: $F = 60$, $D = 20$, target $= 30$.

methods and replacing the REM and the GOTO with a more elegantly structured loop. Some computers support "WHILE TRUE ... WEND", while on others you can use "REPEAT ... UNTIL FALSE". You might even try "FOR T = 0 TO 5 STEP DT ... NEXT T".

Clearly we have succeeded in finding a way to simulate the second-order problem, and there seems no reason why the same approach should not work for third, fourth, fifth, ... How can the approach be formalized?

First we must find a set of variables that describe the present state of the system—in this case x and v. These we will glorify with the title 'state variables'. They must all have derivatives, expressible as combinations of just the variables and the input (or inputs), so that we have a set of equations (with no unknown factors) that express the change of each variable from instant to instant. If there should happen to be some unknown term, then we have clearly left out one of the state variables, and we must hunt for its derivative to work it in as an extra equation.

In the present example, the equations can be laid out as

$$\frac{dx}{dt} = 1 . v$$

$$\frac{dv}{dt} = bu$$

or in the matrix form:

$$\begin{bmatrix} \dfrac{dx}{dt} \\ \dfrac{dv}{dt} \end{bmatrix} = \begin{bmatrix} 0 & 1 \\ 0 & 0 \end{bmatrix} \begin{bmatrix} x \\ v \end{bmatrix} + \begin{bmatrix} 0 \\ b \end{bmatrix} u$$

Admittedly the elaboration appears somewhat useless in this simple example, but we will soon find that system after system falls into the matrix mould of:

$$\dot{\mathbf{x}} = A\mathbf{x} + B\mathbf{u}$$

(Matrix algebra is introduced in the next Prelude.)

2-6 ANALOGUE SIMULATION

It is ironic that analogue simulation 'went out of fashion' just as the solid state operational amplifier was perfected. Previously the integrators had involved a variety of mechanical, hydraulic and pneumatic contraptions, followed by an assortment of electronics based on magnetic amplifiers or thermionic valves. Valve amplifiers were common even in the late sixties, and required elaborate stabilization to overcome their drift. Power consumption was high and air-conditioning essential.

Soon an operational amplifier was available on a single chip, then four to a chip at a price of a few pence, but by then it was deemed easier and more accurate to simulate a system on a digital computer. The cost of analogue computing had become that of achieving tolerances of 0.01 per cent for resistors and capacitors, and of constructing and maintaining the large precision patchboards on which each problem was set up.

In the laboratory, the analogue computer still has its uses. Leave out the patchboard and solder up a simple problem directly. Forget the 0.1 per cent components—the parameters of the system being modelled are probably not known to better than a per cent or two, anyway. Add a potentiometer or two to set up feedback gains, and a great deal of valuable experience can be acquired. Take the problem of the previous section, for example.

An analogue integrator is made from an operational amplifier by connecting the non-inverting input to a common rail, analogue ground, while the amplifier supplies are at $+12$ volts and -12 volts to that rail (more or less). The inverting input is now regarded as a 'summing junction'. A feedback capacitor of, say, 10 microfarads connects the output to this junction, while inputs are each connected via their own 100 kilohm resistor. Such an integrator will have a 'time constant' of one second; if $+1$ volt is applied at one input, the output will change by 1 volt (negatively) in 1 second.

To produce an output that will change in the positive sense, we must follow this integrator with an invertor, a circuit that will give -1 volt output for $+1$ volt input. Another operational amplifier is used, also with the non-inverting input grounded. The feedback to the summing junction now takes the form of a 100 kilohm resistor, and the single input is connected through another 100 kilohm resistor (Fig. 2-5).

The gain of a virtual-earth amplifier is deduced by assuming that the feedback always succeeds in maintaining the summing junction very close to the analogue ground and by assuming that no input current is taken by the operational amplifier. The feedback current of an integrator is then $C \, dv/dt$, where v is the output voltage. The input current is v_{in}/R, and so

$$C\frac{dv}{dt} + \frac{1}{R} v_{in} = 0$$

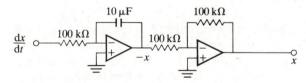

Figure 2-5 Construction of an integrator.

Now if the integrator's 10 microfarad feedback is reduced to 1 microfarad, its time constant will only be 0.1 second. If instead the resistor R is doubled to 200 kilohms the time constant will also be doubled. Various gains can be achieved by selecting appropriate input resistors.

Now in our example we have two 'state variables' x and v, governed by

$$\frac{dx}{dt} = v$$

and

$$\frac{dv}{dt} = bu$$

If we set up two integrators, designating the outputs as x and v, and if we connect the input of the x integrator to the output of the v integrator and the input of the v integrator to the input u through an appropriate resistor, then the simulation is achieved (Fig. 2-6).

As in the case of the digital simulation, the exercise only becomes interesting when we add some feedback. If we connect a potentiometer resistor between the outputs v and $-v$ of the v operational amplifiers, then the wiper can pick off a term of either sign. Thus we can add positive or negative damping. Another potentiometer connected between the supplies can represent the position demanded, while a further one feeds back the position error. The potentiometer signals are mixed in an invertor (which doubles as a summer), giving an output u which is applied to the v input (see Fig. 2-7).

Now the differential state equations merely list the inputs to each integrator in terms of system inputs and the state variables themselves. If we ignore u as such and look at the integrator inputs of our closed-loop system, we see that dx/dt is still v, but now dv/dt has a mixture of inputs. The

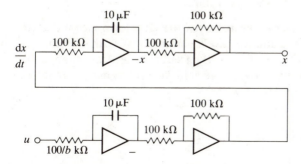

Figure 2-6 Simulating a second-order system.

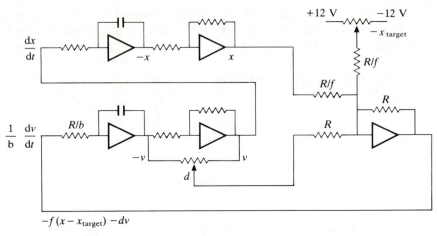

Figure 2-7 Second-order system with variable feedback.

equations have become

$$\frac{dx}{dt} = v$$

and

$$\frac{dv}{dt} = -bfx \quad -bd \quad v \quad + \quad bfx_{target}$$

$$\begin{bmatrix} \dfrac{dx}{dt} \\[2mm] \dfrac{dv}{dt} \end{bmatrix} = \begin{bmatrix} 0 & 1 \\ -bf & -bd \end{bmatrix} \begin{bmatrix} x \\ v \end{bmatrix} + \begin{bmatrix} 0 \\ bf \end{bmatrix} x_{target}$$

We see that applying feedback has turned our second-order system, described by a matrix equation

$$\dot{x} = Ax + Bu$$

into another such system, describable in exactly the same form except that the *A* and *B* matrices are different, and we have a new input. We can perhaps sum up the problem of control as follows: 'By feedback, we can change the matrices describing the system. How do we achieve a set of matrices that we like better than the ones we started with?'

Prelude to THREE

MATRIX ALGEBRA

P3-1 INTRODUCTION

There is a tendency among mathematicians to regard matrices as arcane and mystic entities, with cryptic properties which reward a lifetime of study. For our purposes here they are a convenient shorthand, with properties entirely predictable from the task they perform.

You will probably have first encountered matrices in the solution of simultaneous equations. To take a simple example, the equations:

$$5x + 7y = 2$$

$$2x + 3y = 1$$

can be 'tidied up' by separating the coefficients from the variables in the form:

$$\begin{bmatrix} 5 & 7 \\ 2 & 3 \end{bmatrix} \begin{bmatrix} x \\ y \end{bmatrix} = \begin{bmatrix} 2 \\ 1 \end{bmatrix}$$

Already we see the convention arising that vectors are usually represented by columns, rather than rows. Moreover, the multiplication rule has defined itself, whereby elements taken *across* the matrix are multiplied in turn by elements taken *down* the vector; thus 5 times x is added to 7 times y to give the top element of the product.

P3-2 MATRIX MULTIPLICATION

One way of looking at a matrix is to regard it as defining a mixture of dissimilar units. Let us take a childish example. In a sweetshop, 'Sucks',

19

'Munches' and 'Chews' are on sale. Also on sale are 'Jumbo' bags, each containing 2 Sucks, 3 Munches and 4 Chews, and 'Giant' bags, containing 5 Sucks, 6 Munches and only 1 Chew. If 1 purchase 7 Jumbo bags and 8 Giant bags, how many of each sweet have I bought?

The bag contents can be expressed algebraically as

$$J = 2s + 3m + 4c$$

and
$$G = 5s + 6m + 1c$$

or in matrix form as

$$\begin{bmatrix} J \\ G \end{bmatrix} = \begin{bmatrix} 2 & 3 & 4 \\ 5 & 6 & 1 \end{bmatrix} \begin{bmatrix} s \\ m \\ c \end{bmatrix} \tag{P3-1}$$

Note that matrices do not have to be square, as long as the terms to be multiplied correspond in number.

Now my purchase of 7 Jumbo bags and 8 Giant bags can be written as

$$7J + 8G$$

or in grander form as the product of a row vector with a column vector:

$$\begin{bmatrix} 7 & 8 \end{bmatrix} \begin{bmatrix} J \\ G \end{bmatrix}$$

However, the equations buried in (P3-1) tell me that I can substitute for the J, G vector to obtain

$$\begin{bmatrix} 7 & 8 \end{bmatrix} \begin{bmatrix} 2 & 3 & 4 \\ 5 & 6 & 1 \end{bmatrix} \begin{bmatrix} s \\ m \\ c \end{bmatrix} \tag{P3-2}$$

Now I am faced with the product of a numerical row vector with a numerical matrix. What are the rules? Well, first try common sense. From 7 Jumbo bags, with Sucks at two to a bag, I find 7 times 2 Sucks. From 8 Giant bags I find 8 times 5 more, giving a grand total of 54. So once again the elements *across* the first term are multiplied by elements *down* the columns of the second term to give the result. The answer is, of course,

$$\begin{bmatrix} 54 & 69 & 36 \end{bmatrix} \begin{bmatrix} s \\ m \\ c \end{bmatrix}$$

that is 54 Sucks, 69 Munches and 36 Chews.

Now the shop is selling an Easter bundle of 3 Jumbo bags and a Giant bag, and still has in stock Christmas bundles of 2 Jumbo bags and 4 Giant bags. In no time we can write:

$$\begin{bmatrix} E \\ C \end{bmatrix} = \begin{bmatrix} 3 & 1 \\ 2 & 4 \end{bmatrix} \begin{bmatrix} J \\ G \end{bmatrix}$$

Now if I buy 5 Easter packs and a Christmas pack, I have

$$\begin{bmatrix} 5 & 1 \end{bmatrix} \begin{bmatrix} E \\ C \end{bmatrix} = \begin{bmatrix} 5 & 1 \end{bmatrix} \begin{bmatrix} 3 & 1 \\ 2 & 4 \end{bmatrix} \begin{bmatrix} J \\ G \end{bmatrix}$$

$$= \begin{bmatrix} 5 & 1 \end{bmatrix} \begin{bmatrix} 3 & 1 \\ 2 & 4 \end{bmatrix} \begin{bmatrix} 2 & 3 & 4 \\ 5 & 6 & 1 \end{bmatrix} \begin{bmatrix} s \\ m \\ c \end{bmatrix}$$

$$= \begin{bmatrix} 79 & 105 & 77 \end{bmatrix} \begin{bmatrix} s \\ m \\ c \end{bmatrix}$$

$$= 79 \text{ Sucks} + 105 \text{ Munches} + 77 \text{ Chews (prove it!)}$$

All that is needed in principle is to keep multiplying the matrices by the rules, which have become obvious.

The mathematician will still worry about the order in which the matrix multiplication is carried out. Clearly the order of the matrices cannot be jumbled. If we swap the terms of (P3-2), we have

$$\begin{bmatrix} 2 & 3 & 4 \\ 5 & 6 & 1 \end{bmatrix} \begin{bmatrix} 7 & 8 \end{bmatrix} \begin{bmatrix} s \\ m \\ c \end{bmatrix}$$

But now nothing fits. If we multiply the last two terms, we have $7s + 8m$—but what is there to multiply by c? Multiply the first two terms, to get 2 times 7, but the 3 and the 4 are hopelessly lost. The order of the terms must be carefully preserved.

On the other hand, the matrix multiplication task may be tackled from either end. Taking our sticky example, the Christmas and Easter bags can first be opened to reveal a total of Jumbo and Giant bags, then these can be expanded into individual sweets, or alternatively the total of each sweet for a Christmas bag and for an Easter bag can be worked out first; the result must be the same.

P3-3 TRANSPOSITION OF MATRICES

Our sweetshop arithmetic has featured column vectors representing the types of object—Munches, Chews, Jumbo bags, etc.—while the quantities have been in the form of row vectors. When the product is a scalar, we could clearly change the order. Now

$$[96 \quad 88 \quad 77] \begin{bmatrix} s \\ m \\ c \end{bmatrix}$$

clearly gives the same result as

$$[s \quad m \quad c] \begin{bmatrix} 96 \\ 88 \\ 77 \end{bmatrix}$$

The column vector in the second example is the *transpose* of the row vector [96   88   77], and transposition is denoted with a dash: [96   88   77]'. The matrix is in general reflected about a diagonal line top-left to bottom-right. All the arguments of the last section can be pursued with the transpose of each matrix, but we see an important difference.

Equation (P3-1) can be rearranged to give

$$[J \quad G] = [s \quad m \quad c] \begin{bmatrix} 2 & 5 \\ 3 & 6 \\ 4 & 1 \end{bmatrix}$$

so our purchase of 7 Jumbo bags and 8 Giant bags now expands as

$$[J \quad G]\begin{bmatrix} 7 \\ 8 \end{bmatrix} = [s \quad m \quad c] \begin{bmatrix} 2 & 5 \\ 3 & 6 \\ 4 & 1 \end{bmatrix}\begin{bmatrix} 7 \\ 8 \end{bmatrix}$$

while the gluttony of 5 Easter packs and a Christmas pack can be written as

$$[E \quad C]\begin{bmatrix} 5 \\ 1 \end{bmatrix} = [s \quad m \quad c] \begin{bmatrix} 2 & 5 \\ 3 & 6 \\ 4 & 1 \end{bmatrix}\begin{bmatrix} 3 & 2 & 5 \\ 1 & 4 & 1 \end{bmatrix}$$

Note that not only are all the matrices transposed but the order in which they appear is reversed. We have the important generalization that if A, B and C are matrices, then

$$ABC = (AB)C = A(BC) \tag{P3-3}$$

(the matrix product is *associative*) and

$$(ABC)' = C'B'A' \tag{P3-4}$$

Moreover, for the product to have any meaning, the number of columns of A must equal the number of rows of B, while the number of columns of B must equal the number of rows of C. The result will have as many rows as A and as many columns as C.

One last point to note before moving on is that

$$\begin{bmatrix} 1 & 0 & 0 \\ 0 & 1 & 0 \\ 0 & 0 & 1 \end{bmatrix} \begin{bmatrix} s \\ m \\ c \end{bmatrix} = \begin{bmatrix} s \\ m \\ c \end{bmatrix}$$

The matrix on the left is the 3 by 3 *unit matrix*. If placed before any matrix with three rows, the product will be the same as the original matrix. Note, too, that if it is placed after any matrix with three columns, the product again produces an unchanged result. The unit matrix may not seem to achieve a lot, but it will come in very useful later. It is usually denoted by the letter I. Its actual size may differ from 3 by 3—it is assumed to suit the occasion.

P3-4 COORDINATE TRANSFORMATIONS

Vector geometry is usually introduced with the aid of three orthogonal unit vectors **i**, **j** and **k**. It is explained that the vector (x, y, z) is really a mixture of these unit vectors:

$$x\mathbf{i} + y\mathbf{j} + z\mathbf{k}$$

This expansion can be represented in the smarter form:

$$[\mathbf{i} \quad \mathbf{j} \quad \mathbf{k}] \begin{bmatrix} x \\ y \\ z \end{bmatrix}$$

For now, let us keep to two dimensions and consider just (x, y).

Suppose that there are two sets of axes in action. With respect to our first set the point is (x, y) but with respect to a second set it is (u, v). Just how are these two vectors related?

What we have in effect is one pair of unit vectors **i** and **j**, and another pair **l** and **m**, say. Since both sets of coordinates represent the same vector, we have

$$x\mathbf{i} + y\mathbf{j} = u\mathbf{l} + v\mathbf{m}$$

that is

$$[\mathbf{i} \quad \mathbf{j}] \begin{bmatrix} x \\ y \end{bmatrix} = [\mathbf{l} \quad \mathbf{m}] \begin{bmatrix} u \\ v \end{bmatrix} \qquad \text{(P3-5)}$$

Now each of the vectors **l** and **m** must be expressible in terms of **i** and **j**. We may discover that

$$\mathbf{l} = a\mathbf{i} + b\mathbf{j}$$

and

$$\mathbf{m} = c\mathbf{i} + d\mathbf{j}$$

or in matrix form:

$$[1 \quad m] = [i \quad j]\begin{bmatrix} a & c \\ b & d \end{bmatrix} \tag{P3-6}$$

We want the relationship in this slightly twisted form because we want to substitute into (P3-5) to eliminate one pair of vectors:

$$[i \quad j]\begin{bmatrix} x \\ y \end{bmatrix} = [i \quad j]\begin{bmatrix} a & c \\ b & d \end{bmatrix}\begin{bmatrix} u \\ v \end{bmatrix} \tag{P3-7}$$

Now Eq. (P3-7) declares that two mixtures of the unit vectors are equal; unless the unit vectors are in the same direction (and they are not) the ingredients must match, that is

$$\begin{bmatrix} x \\ y \end{bmatrix} = \begin{bmatrix} a & c \\ b & d \end{bmatrix}\begin{bmatrix} u \\ v \end{bmatrix} \tag{P3-8}$$

Although this exercise is now graced with the name *vector geometry*, we are merely adding up mixtures in just the same form as the antics in the sweetshop. To convert our (u, v) coordinates into the (x, y) frame, we simply multiply the coordinates by an appropriate matrix that defines the mixture.

Suppose, however, we are presented with the value(s) of (x, y) and are asked to find (u, v). We are left trying to solve two simultaneous equations:

$$x = au + cv$$

and

$$y = bu + dv$$

In traditional style we multiply the top equation by d and subtract c times the second equation to obtain

$$dx - cy = (ad - bc)u$$

and in a similar way we find

$$-bx + ay = (ad - bc)v$$

which we can rearrange as

$$\begin{bmatrix} u \\ v \end{bmatrix} = \frac{1}{ad - bc}\begin{bmatrix} d & -c \\ -b & a \end{bmatrix}\begin{bmatrix} x \\ y \end{bmatrix}$$

where the constant $1/(ad - bc)$ multiplies each of the coefficients inside the matrix. If the original relationship between $(x, y)'$ and $(u, v)'$ was

$$\begin{bmatrix} x \\ y \end{bmatrix} = T\begin{bmatrix} u \\ v \end{bmatrix}$$

then we have found an *inverse matrix* such that

$$\begin{bmatrix} u \\ v \end{bmatrix} = T^{-1} \begin{bmatrix} x \\ y \end{bmatrix}$$

The value of $(ad - bc)$ obviously has special importance—we will have great trouble in finding an inverse if $(ad - bc) = 0$. Its value is the *determinant* of the matrix T.

P3-5 MATRICES, NOTATION AND COMPUTING

Rather than use separate variables x, y, u, v and so on, it is more convenient mathematically to use *subscripted variables* as the elements of a vector. The entire vector is then represented by the single symbol **x**, which will be made up of elements x_1, x_2 and so on. Matrices are now made up of elements with two suffices; thus

$$A = \begin{bmatrix} a_{11} & a_{12} & a_{13} \\ a_{21} & a_{22} & a_{23} \\ a_{31} & a_{32} & a_{33} \end{bmatrix}$$

In a computer program, the subscripts appear in brackets, so that a vector could be represented by the elements X(1), X(2) and X(3), while the elements of the matrix are A(1, 1), A(1, 2) and so on. It is in matrix operations that this notation really earns its keep. Suppose that we have a relationship

$$\mathbf{x} = T\mathbf{u}$$

where the vectors have three elements and the matrix is 3 by 3. Instead of a massive block of arithmetic, the entire product is expressed in just five lines of program:

```
FOR I = 1 TO 3
X(I) = 0
FOR J = 1 TO 3
X(I) = X(I) + T(I, J)*U(J)
NEXT J, I
```

For the matrix product $C = AB$ the program is hardly any more complex:

```
FOR I = 1 TO 3
FOR J = 1 TO 3
C(I, J) = 0
FOR K = 1 TO 3
C(I, J) = C(I, J) + A(I, K)*B(K, J)
NEXT K, J, I
```

Although these examples are illustrated in BASIC, they would look almost identical in a variety of languages and would show the same economy of programming effort. Clearly, if we are to try to analyse any but the simplest of systems by computer, we should first represent the problem in a matrix state equation form.

THREE

ADDING CONTROL

3-1 INTRODUCTION

Some approaches to control theory draw a magical boundary between open loop and closed loop systems. Yet towards the end of Chapter 2 we saw that a similar looking set of state equations described either open or closed loop behaviour. Our problem is to find out how to modify these equations by means of feedback, so that the system in its new form behaves in a way that is more desirable than it was before.

3-2 VECTOR STATE EQUATIONS

It is tempting to assert that every dynamic system can be represented by a set of state equations in the form:

$$\dot{\mathbf{x}} = A\mathbf{x} + B\mathbf{u} \tag{3-1}$$

Unfortunately there are many exceptions. A sharp switching action cannot reasonably be expressed by differential equations of any sort. A highly non-linear system will only approximate to the above form of equations over a small disturbance. A reel of magnetic tape will have a variable for each fragment of ferrite, and again cannot reasonably be expressed as above. Nevertheless, the majority of systems with which the control engineer is concerned fall closely enough into this form that it becomes a very useful tool indeed.

For all its virtues, in describing the way in which the state of the system

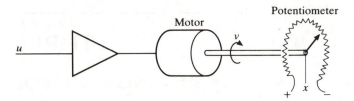

Figure 3-1 Position controller.

changes with time, Eq. (3-1) is only half of the story. Suppose that we look again at the motor position controller, with a potentiometer to measure the output position and some other constants defined for good measure (Fig. 3-1).

Once again we have state variables x and v, representing the position and velocity of the output. If the input drive is u, we may more neatly represent the state variables as x_1 (for x) and x_2 (for v) and write the differential equations in matrix form:

$$\begin{bmatrix} \dot{x}_1 \\ \dot{x}_2 \end{bmatrix} = \begin{bmatrix} 0 & 1 \\ 0 & -a \end{bmatrix} \begin{bmatrix} x_1 \\ x_2 \end{bmatrix} + \begin{bmatrix} 0 \\ b \end{bmatrix} u \qquad (3\text{-}2)$$

For feedback purposes, what concerns us is the output of the potentiometer, y. In this case, we can write $y = cx$, but to be more general we should regard this as a special case of a matrix equation:

$$\mathbf{y} = C\mathbf{x} \qquad (3\text{-}3)$$

In the present case, we can only measure the position; we may be able to guess at the velocity, but without adding extra filtering circuitry we cannot feed it back. If, on the other hand, we had a tacho to measure the velocity directly, then the output $\mathbf{y}$ would become a vector with two components, one proportional to position and the other proportional to the velocity.

For that matter, we could add two tachos to obtain three output signals, and tack on a few more potentiometers into the bargain. They would be of little help in controlling the system, but the point is that the number of outputs is simply the number of sensors, which may be none (not much hope

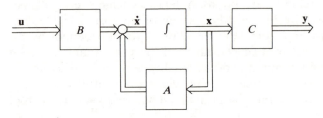

Figure 3-2 System equations in block diagram form.

for control!), one, less than, equal to or more than the number of state variables. Futile as it might appear at first glance, adding extra sensors has a purpose when making a system such as an autopilot, where the control system must be able to survive the loss of one or more signals. The system equations are shown in block diagram form in Fig. 3-2.

3-3 FEEDBACK

The input to our system is at present the vector **u**, here only having one component. To apply feedback, we must mix proportions of our output signals with a command input, **v**, to construct the input **u** which we apply to the system (see Fig. 3-3).

Now

$$\mathbf{u} = F\mathbf{y} + G\mathbf{v}$$

When we substitute this into Eq. (3-1) we obtain

$$\dot{\mathbf{x}} = A\mathbf{x} + B(F\mathbf{y} + G\mathbf{v}) \tag{3-4}$$

However, (3-3) tells us that

$$\mathbf{y} = C\mathbf{x}$$

from which we deduce that

$$\dot{\mathbf{x}} = (A + BFC)\mathbf{x} + BG\mathbf{v} \tag{3-5}$$

If B, the input matrix, and C, the output matrix, both had four useful elements then we could choose the four components of F to achieve any closed loop state matrix we wished. Unfortunately in this case they have degenerated to

$$B = \begin{bmatrix} 0 \\ b \end{bmatrix} \quad \text{and} \quad C = [c \quad 0]$$

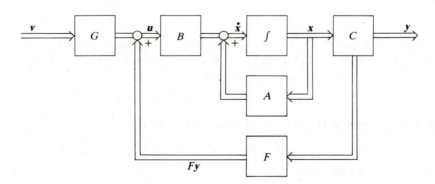

Figure 3-3 A system with feedback.

so that F has reduced to a single scalar f. We see that

$$BFC = \begin{bmatrix} 0 & 0 \\ bfc & 0 \end{bmatrix}$$

so that

$$A + BFC = \begin{bmatrix} 0 & 1 \\ bfc & -a \end{bmatrix}$$

3-4 ANOTHER APPROACH

The position control example can be attacked in the 'traditional' way as follows. The shaft position is subject to a differential equation deduced from simple mechanics. The applied torque contributes to the acceleration of the system's inertia and also overcomes the damping effect of the velocity drag to give

$$ku = I\ddot{x} + h\dot{x}$$

where k relates the input voltage to the torque, I is the inertia of the system and h is the damping torque per unit of speed. These are rearranged to give

$$\ddot{x} + a\dot{x} = bu \tag{3-6}$$

where $a = h/I$, $b = k/I$. This equation corresponds directly with Eq. (3-2).

The potentiometer gives an output y proportional to x; let us say $y = cx$. Feedback means mixing this potentiometer signal with another input v (do not confuse it with the velocity symbol we used before) and applying the result to the input so that $u = fy + gv$. Substituting this into (3-6) we get

$$\ddot{x} + a\dot{x} = b(fcx + gv)$$

that is

$$\ddot{x} + a\dot{x} - bfcx = bgv \tag{3-7}$$

It all looks very familiar. Let us put some numbers to the constants. Suppose that $a = 5$, $b = 6$ and $c = 1$ volt per unit. Suppose also that we have chosen to set $f = -1$ and $g = 1$. Then (3-7) becomes

$$\ddot{x} + 5\dot{x} + 6x = 6v$$

If we use the symbol D to represent d/dt, we can write this as

$$(D^2 + 5D + 6)x = 6v$$

which can be factorized as

$$(D + 2)(D + 3)x = 6v$$

Now $(D + 3)x$ is the same as $\dot{x} + 3x$. If we define a new variable, w_1, to be equal to $\dot{x} + 3x$ then we have

$$(D + 2)w_1 = 6v$$

or
$$\dot{w}_1 = -2w_1 + 6v$$

Similarly, if $w_2 = \dot{x} + 2x$ then

$$(D + 3)w_2 = 6v$$

or
$$\dot{w}_2 = -3w_2 + 6v$$

Suddenly we have two new state equations, expressible in matrix form as

$$\dot{\mathbf{w}} = \begin{bmatrix} -2 & 0 \\ 0 & -3 \end{bmatrix} \mathbf{w} + \begin{bmatrix} 6 \\ 6 \end{bmatrix} v$$

These new variables are expressed in terms of the old ones by

$$w_1 = 3x_1 + x_2$$

and
$$w_2 = 2x_1 + x_2$$

or in matrix form as

$$\mathbf{w} = \begin{bmatrix} 3 & 1 \\ 2 & 1 \end{bmatrix} \mathbf{x}$$

We also need to find the value of the output, y, and we see that it is given by

$$y = \begin{bmatrix} 1 & -1 \end{bmatrix} \mathbf{w}$$

These new equations are somehow superior to the old ones, in that the two w's are independent of each other and can be solved as completely separate first-order problems.

Let us recap by expressing these latest manoeuvres (after closing the feedback loop) in matrix form. We started with a system describable in the form:

$$\dot{\mathbf{x}} = A\mathbf{x} + B\mathbf{u}$$

with output given by

$$\mathbf{y} = C\mathbf{x}$$

We then found new variables $\mathbf{w}$ given by a linear transformation

$$\mathbf{w} = T\mathbf{x}$$

that is
$$\mathbf{x} = T^{-1}\mathbf{w}$$

The rates of change of these variables will be

$$\dot{\mathbf{w}} = T\dot{\mathbf{x}}$$
$$= T(A\mathbf{x} + B\mathbf{u})$$
$$= TAT^{-1}\mathbf{w} + TB\mathbf{u}$$

Meanwhile,

$$\mathbf{y} = C\mathbf{x}$$
$$= CT^{-1}\mathbf{w}$$

Given any one set of state variables, we can produce a new set of state equations in terms of a transformation of them. The variety of equations is literally infinite for any one problem. However, just one or two forms will be of particular interest, as we will see later.

3-5 SYSTEMS WITH REAL COMPONENTS—USE OF THE PHASE PLANE

Linear theory assumes that equations hold true over any range of values. If the position error and velocity of the position control system are doubled, then the drive will also double. Double them again and again, and the drive can be increased without limit. Common sense tells us that this is an oversimplified view of the real world; sooner or later the drive signal will reach a maximum value beyond which it cannot go.

To deal with second-order problems such as this, there is a useful graphical technique, the phase plane. Phase plane and state space are in fact one and the same; the only difference here is that we are concerned with plotting *trajectories* of the state in the form of a graph of velocity versus position. First let us explore this technique in the linear case, before the motor starts to limit.

We will stick with the now familiar example, a servo system defined by the second-order differential equation

$$\ddot{x} + 5\dot{x} + 6x = 6v$$

where v is an input value of position demand. To assess the performance of the control system, it is sufficient to set the demanded position to zero and to see how the system responds to a disturbance, i.e. to some initial value of position and velocity. Now we can rearrange the equation to set the acceleration on the left equal to feedback terms on the right:

$$\ddot{x} = -6x - 5\dot{x} \tag{3-8}$$

If we are to trace the track of the (position, velocity) coordinates around the phase plane, it would be helpful to know in which direction they might move. Of particular interest is the slope of their curve at any point,

$$\frac{d(\dot{x})}{dx}$$

We wish to look at the derivative of $\dot{x}$ with respect to x, not with respect to time as we usually do. These derivatives are closely related, however, since for the general function f,

$$\frac{df}{dx} = \frac{dt}{dx}\frac{df}{dt} = \frac{1}{\dot{x}}\frac{df}{dt}$$

So we have

$$\text{Slope} = \frac{d\dot{x}}{dx} = \frac{1}{\dot{x}}\ddot{x} \tag{3-9}$$

In this case, $\ddot{x}$ is expressed by Eq. (3-8), and so the slope is given by

$$\text{Slope} = -6\frac{x}{\dot{x}} - 5 \tag{3-10}$$

The first thing we notice is that for all points where x and $\dot{x}$ are in the same ratio, the slope is the same. The lines of constant ratio all pass through the origin. On the line $x = 0$, we have slope -5. On the line $\dot{x} = x$ we have slope -11. On the line $\dot{x} = -x$ we have slope $+1$, and so on. We can make up a spider's web of lines, with small dashes showing the directions in which the trajectories will cross them. These lines on which the slope is the same are termed *isoclines*.

An interesting isocline is the line $6x + 5\dot{x} = 0$. On this line the acceleration is zero and so the isocline represents points of zero slope; the trajectories cross the line horizontally.

In this particular example, there are two isoclines that are worth closer scrutiny. Consider the line $\dot{x} = -2x$. Here the slope is -2—exactly the same as the slope of the line itself. Once the trajectory encounters this line, it will lock on and never leave it. The same is true for the line $\dot{x} = -3x$, where the trajectory slope is found to be -3.

Is it a coincidence that the 'special' state variables found in Sec. 3-4 could be expressed as $\dot{x} + 2x$ and $\dot{x} + 3x$ respectively?

Having mapped out the isoclines, we can steer around the plane, following the local slope to map out sets of trajectories. This is shown in Fig. 3-4.

The phase plane 'portrait' gives a good insight into the system's behaviour, without having to make any attempt to solve its equations. We see that for any starting point, the trajectory homes in on one of the special isoclines and settles without any oscillatory behaviour. As an exercise, sketch

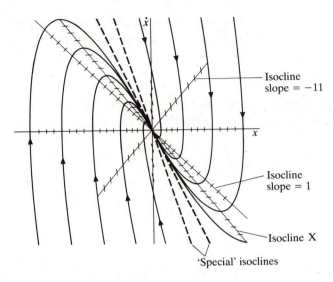

Figure 3-4 Phase plane diagram showing isoclines.

the phase plane for the system

$$\ddot{x} + \dot{x} + 6x = 0$$

i.e. for the same position control system, but with much reduced velocity feedback. You will find an equation similar to (3-10) for the slopes on the isoclines, but will not find any 'special' isoclines. The trajectories will match the 'spider's web' image better than before, spiralling in to the origin to represent system responses which now are lightly damped sine waves.

Turning back to the example of Fig. 3-4 we see that the acceleration is given by Eq. (3-8), and if we introduce the value of the drive, u, the equation becomes

$$\ddot{x} = u$$

where

$$u = -6x - 5\dot{x}$$

In a practical system, the drive amplifier will saturate if the signal fed back from the error becomes too large. How can we take account of this? By way of an example, let us assume that the motor relies entirely on feedback for its damping, and thus that the acceleration is entirely the result of the drive amplifier. Let us also assume that the drive signal u is limited by

$$|u| < 4$$

On the line

$$6x + 5\dot{x} = 0$$

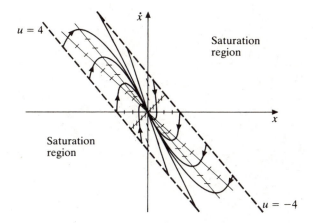

Figure 3-5 Phase plane of linear region where $|u| < 4$.

we have noted that the acceleration is zero because the drive is zero. On the parallel line

$$6x + 5\dot{x} = 2$$

the acceleration will have the value -2. As we move to lines further and further from the zero drive line the acceleration will increase in magnitude, until we reach the line, say at an acceleration value of -4, where the drive saturates. On the other side of the zero drive line, at an acceleration of $+4$, the drive will saturate in the opposite sense.

We thus have two parallel lines within which the behaviour is linear and where our phase plane portrait will tell the truth. Outside these lines, however, the drive will saturate and the system's behaviour will no longer be as before (see Fig. 3-5).

To fill in the mystery region in this phase plane, we must find out how the system will behave under saturated drive.

Equation (3-9) tells us that the slope of the trajectory is always given by $\ddot{x}/\dot{x}$, so in this case we are interested in the case where $\ddot{x}$ has saturated at a value of $+4$ or -4, giving the slope of the trajectories as $4/\dot{x}$ or $-4/\dot{x}$.

We can see that the isoclines are no longer lines through the origin as before, but are lines of constant $\dot{x}$, parallel to the horizontal axis. If we want to find out the actual shape of the trajectories, we must solve for $\dot{x}$ in terms of x. We can integrate the expression for $\ddot{x}$ twice to see

$$\ddot{x} = 4$$

so

$$\dot{x} = 4t + a \tag{3-11}$$

and

$$x = 2t^2 + at + b \tag{3-12}$$

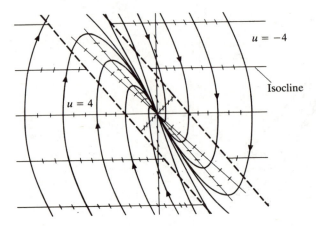

Figure 3-6 Outside the linear region $u = 4$ or -4. The phase plane is completed with the aid of isoclines parallel to the x axis.

Squaring (3-11) and subtracting eight times (3-12) reveals that

$$\dot{x}^2 - 8x = a^2 - 8b$$

i.e. the trajectories are parabolae of the form

$$\dot{x}^2 = 8x + c$$

or

$$\dot{x}^2 = -8x + c$$

for the trajectories with drive of value -4. The three regions of the phase plane can now be cut and pasted together to give the full picture of Fig. 3-6.

The phase plane can become a powerful tool for the design of high-performance position control systems. The motor drive might be proportional to the position error for small errors, but to achieve accuracy the drive must approach its limit for quite a small displacement. The 'proportional band' is small, and for any substantial disturbance the drive will spend much of its time saturated. The ability to design feedback on a non-linear basis is then of great importance.

For now, try the following exercise. (A solution is given in Appendix B.)

Exercise 3-5 *The position control system is described by the same equations as before:*

$$\ddot{x} + 5\dot{x} + 6x = 0$$

This time, however, the damping term $5\dot{x}$ is given not by feedback but by passive damping in the motor, that is

$$\ddot{x} = -5\dot{x} + u$$

where

$$u = -6x$$

Once again the drive, u, saturates at values $+4$ or -4. Sketch the new phase plane, noting that the saturated trajectories will no longer be parabolae.

3-6 MORE NON-LINEARITIES

In an electronic controller, a sharp non-linearity can occur as an amplifier saturates. In the world at large, we are lucky to find any system that is truly linear. The tension in a spring is proportional to its extension—more or less. Rate of loss of heat is proportional to temperature—provided the temperature does not vary too much. The expression of a system in a linear form

$$\dot{\mathbf{x}} = A\mathbf{x} + B\mathbf{u}$$

is nearly always only an approximation to the truth. Strictly we should regard this as a special case of

$$\dot{\mathbf{x}} = \mathbf{f}(\mathbf{x}, \mathbf{u})$$

For small perturbations, we can make a linear approximation to f of the form

$$\mathbf{f}(\mathbf{x}_0 + \delta x, \mathbf{u}_0 + \delta u) = f(x_0, u_0) + a\,\delta x_1 + b\,\delta x_2 + \cdots + p\,\delta u_1 + q\,\delta u_2 + \cdots$$

which brings us back to the linear equations if we let our state variables be the small perturbations $\delta\mathbf{x}$. There is the matter of the embarrassing vector of constants $f(x_0, u_0)$. However, we can work these in by adding another component to the input, having constant value 1, and adding an extra column to the matrix B to share out the appropriate constants among the derivatives $\dot{\mathbf{x}}$.

We can see a constant offset in action most easily if we look at the classic mass–spring system, as shown in Fig. 3-7. Here a mass M is suspended by a spring which has tension k times its extension. The top of the spring is moved by the input, u, so that the extension is $u - y$, where y is the vertical position of the mass.

Normally we would measure y from the rest position of the mass, and deduce that the mass times the acceleration could be equated to the tension in the spring, that is

$$M\ddot{y} = k(u - y) \tag{3-13}$$

We make the obvious choice of $x_1 = y$, $x_2 = \dot{y}$, and arrive at matrix state equations

$$\begin{bmatrix} \dot{x}_1 \\ \dot{x}_2 \end{bmatrix} = \begin{bmatrix} 0 & 1 \\ -k/M & 0 \end{bmatrix} \begin{bmatrix} x_1 \\ x_2 \end{bmatrix} + \begin{bmatrix} 0 \\ k/M \end{bmatrix} u$$

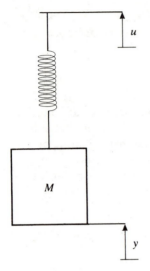

Figure 3-7 Mass M on spring of stiffness k.

If, instead, we measure y from the unstretched position of the spring, we must add the effect of gravity to (3-13) to obtain

$$M\ddot{y} = k(u - y) - Mg \qquad (3\text{-}14)$$

Now the extra constant acceleration $-g$ can be accommodated by adding an extra element 1 to the input, giving

$$\begin{bmatrix} \dot{x}_1 \\ \dot{x}_2 \end{bmatrix} = \begin{bmatrix} 0 & 1 \\ -k/M & 0 \end{bmatrix} \begin{bmatrix} x_1 \\ x_2 \end{bmatrix} + \begin{bmatrix} 0 & 0 \\ k/M & -g \end{bmatrix} \begin{bmatrix} u \\ 1 \end{bmatrix}$$

Let us return to the central problem of non-linearities. Local lineariz-ation is all very well if we expect the disturbances to be small, but that will often not be the case. The phase plane has been seen to be useful in examining piecewise linear systems, and in some cases it is no doubt possible to find isoclines for more general non-linearities. However, we would like to find a method for analysing the stability of non-linear systems in general, including systems of higher order than two.

One approach is the 'direct' method of Liapunov, astonishingly simple in principle but sometimes needing ingenuity to apply. First, how should we define stability?

If we disturb the system, its state will follow a trajectory in n-dimensional state space. If all such trajectories lead back to a single point at which the system comes to rest, then the system is asymptotically stable. If some trajectories diverge to infinity, then the system is unstable.

There is a third possibility. If all trajectories lead to a bounded region of the state space, remaining thereafter within that region without necessarily settling, then the system is said to have 'bounded stability'.

These definitions suggest that we should examine the trajectories to see whether they lead 'inwards' or 'outwards'—whatever that might mean. Suppose that we define a function of the state, $L(\mathbf{x})$, so that the equation

$$L(\mathbf{x}) = r$$

defines a closed 'shell'. (Think of the example of circles or spheres of radius r.) Suppose that the shell for a smaller value of r is totally enclosed in the shell for any larger value of r and that as r is reduced to zero so the shells converge to a single point of the state space, $\mathbf{x}_0$. If we can show that on any trajectory the value of r continuously decreases until r becomes zero, then clearly all trajectories must converge to $\mathbf{x}_0$. The system is asymptotically stable.

Alternatively, if we can find such a function for which r increases indefinitely, then the system is unstable. The skill lies in spotting the function L.

Let us look at a very simple linear example, for which the sophistication of this method is really not needed. Let the system to be investigated be

$$\ddot{x} + \dot{x} + x = 0 \tag{3-15}$$

As the Liapunov function, consider

$$\dot{x}^2 + \dot{x}x + x^2 = r \tag{3-16}$$

To find dr/dt we must differentiate this, giving

$$\frac{dr}{dt} = 2\ddot{x}\dot{x} + \ddot{x}x + \dot{x}^2 + 2\dot{x}x$$

$$= \ddot{x}(2\dot{x} + x) + \dot{x}^2 + 2\dot{x}x$$

We can substitute for $\ddot{x}$ from (3-15) to obtain

$$\frac{dr}{dt} = (-\dot{x} - x)(2\dot{x} + x) + \dot{x}^2 + 2\dot{x}x$$

$$= -\dot{x}^2 - \dot{x}x - x^2$$

$$= -r$$

Clearly on any shell where $r > 0$, the trajectory moves inwards towards the centre and the system is asymptotically stable.

Exercise 3-6-1 *Find a suitable Liapunov function to analyse the more interesting system*

$$\ddot{x} + \dot{x}(\dot{x}^2 + x^2 - 1) + x = 0$$

If we try $r = \dot{x}^2 + x^2$ as a Liapunov function, we find that

$$\frac{dr}{dt} = 2\ddot{x}\dot{x} + 2\dot{x}x$$

$$= 2[-\dot{x}(\dot{x}^2 + x^2 - 1) - x]\dot{x} + 2\dot{x}x$$

$$= -2\dot{x}^2(r - 1)$$

In $r < 1$, $dr/dt \geqslant 0$ so the origin is unstable.

If $r > 1$, $dr/dt \leqslant 0$ so the system has bounded stability.

On $r = 1$, $dr/dt = 0$ so there is a clearly defined limit cycle.

FOUR

PRACTICAL CONTROL SYSTEMS

4-1 INTRODUCTION

An electric iron manages to achieve temperature control with one single bimetal switch. The space shuttle requires somewhat more control complexity. Control systems can be vast or small, can aim at smooth stability or a switching limit cycle, can be designed for supreme performance or can be the cheapest and most expedient way to control a throwaway consumer product. So where should the design of a controller begin?

There must first be some specification of the performance required of the controlled system. In the now-familiar servomotor example, we must be told how accurately the output position must be held, what forces might disturb it, how fast and with what acceleration the output position is required to move. Considerations of reliability and lifetime must then be taken into account. Will the system be required to work only in isolation or is it part of a more complex whole?

A simple radio-controlled model servomotor will use a small d.c. motor, with a potentiometer to measure output position. For small deviations from the target position the amplifier in the loop will apply a voltage to the motor proportional to error, and with luck the output will achieve the desired position without too much overshoot.

An industrial robot arm requires a little more attention. The motor may still be d.c., but will probably be of high performance at no small cost. To ensure a well-damped response, the motor may well have a built-in tachometer which gives a measure of its speed. A potentiometer is hardly good enough, in terms of accuracy or lifespan; an incremental optical transducer is much more likely—although some systems have both. The control loop is not likely to be closed merely by a simple amplifier; a computer is almost certainly needed. Once this level of complexity is reached, position control examples show many common features.

When it comes to the computer that applies the control, it is the control strategy that counts rather than the size of the system. A radio-telescope in the South of England used to be controlled by two mainframes, with dubious success. They were replaced by two personal microcomputers, with a purchase cost of only one twenty-fifth of the mainframes' annual maintenance cost, and the performance is now much improved.

4-2 AN EXPERIMENTAL ROBOT AXIS CONTROLLER

As part of an undergraduate control course, a simplified 'arm' has been developed with shoulder and elbow joints acting in the same plane. With the addition of a further vertical movement and a twisting gripper, the system could be expanded into the 'SCARA' configuration of robot. The d.c. motors have built-in two-phase optical transducers, which with the aid of two 74LS2000 chips keep track of the shoulder and elbow positions as 16-bit numbers. Two D/A converters provide speed command signals, which are mixed into velocity control loops (see Fig. 4-1).

You will have gathered by now that a computer comes into the picture. The experiment was originally developed around the use of an Acorn 'BBC' computer, but has been adapted here to use the more widely available IBM-PC. Whereas the BBC has a buffered bus for attaching special I/O devices, it is more conveneint to interface the hardware to the IBM by means of a *prototype card*. This provides buffered connections to the machine's internal bus, allowing board space for special circuitry.

The system possesses many of the interesting features of a more general on-line computer controller, and is worth a closer look.

4-3 CONNECTING TO THE COMPUTER

A control computer can vary in size from a mainframe to a single-chip micro. For any but the simplest of systems, the 'recipe book' designer will point out that most computer operating systems already support serial communication over standard RS-232 lines and will deduce that the easiest answer is an 'intelligent peripheral'. This has not answered the problem at all; it has merely shifted the task from one of interfacing the master computer to that of interfacing the microcomputer that makes the peripheral intelligent.

Except in the case of some bizarre experimental machines, input/output can be based on the concept of *memory mapping*. Within the computer are

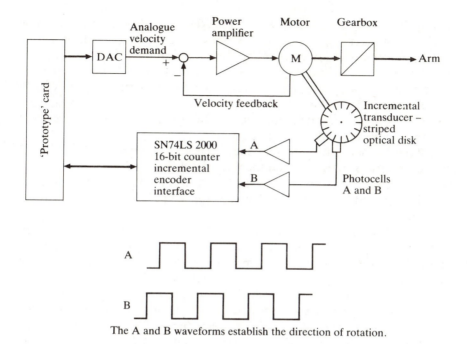

The A and B waveforms establish the direction of rotation.

Figure 4-1 The 'SCARA' position control experiment. The drawings show one of the two motor position control channels of the 'SCARA' experiment. The motors are Portescap 23L11213E-B40Y-1, 12V d.c. motors with integrated 40-stripe optical encoders, fitted to a 30:1 ratio gearbox size A42.

two main *busses*, the address bus and the data bus. If a word needs to be fetched from memory, the address of the word in question is set up by the CPU (central processing unit) on the lines of the address bus. The read/not-write control line is held high, and within a fraction of a microsecond the appropriate memory chip has pulled the bits of the data bus high or low respectively to correspond to the word memorized in the selected address. The CPU will probably copy the data bus to one of its internal registers and will then move on to the next instruction.

What has happened is that, by the selection of a suitable address, a set of minute electrical charges within the memory chip have been tested and the values they represent have been transferred to the CPU. Cannot the same principle be used to read a set of voltages applied to input lines? The answer is, of course, yes.

Specialized peripheral chips are available for a variety of purposes. They appear to the computer as very small memory chips of only a few bytes capacity. However, these bytes represent the status of input lines, or of counters, of received serial signals or of analogue quantities. Just as values can be input to the CPU from such chips, so other values can be output as though the CPU were commanding the interface chip to store a value in its memory, while in fact the chip controls functions in the outside world.

The first stage in reading a memory chip is *address decoding*. The numerous identical memory chips each hold part of the computer's *address space*, and the correct chip must be selected to perform its task. Bits from the most significant part of the address are applied to decoders, so that (say) one of sixteen select lines will be pulled down according to the top 'nibble' of the address.

When one of these slices of space is to be used for interfacing, it will probably be divided down again and again by decoding lower significant parts of the address, leaving only the bottom two or four bits to select registers within the control chip.

Microprocessors of the 8086 family discriminate between memory operations and input/output. There is an additional line on the system but that decides the difference, while for some input/output operations only the bottom eight address bits have relevance. This really makes no difference to the principle, since this extra line can be regarded as just another address bit. Some extra care might be necessary to avoid trampling on the more limited address space of the peripherals normally connected to the machine.

In the IBM-PC and XT (and probably several others in the range) the input/output hexadecimal addresses 300 to 31F are assigned to the prototype interface card. Although not the cheapest interface, the prototype card makes an easy starting point.

On the card there is address decoding. If the value on the computer's address bus falls in the range &H300 to &H31F while I/O read or I/O write are active, then a logic output becomes low, enabling a data buffer to link the

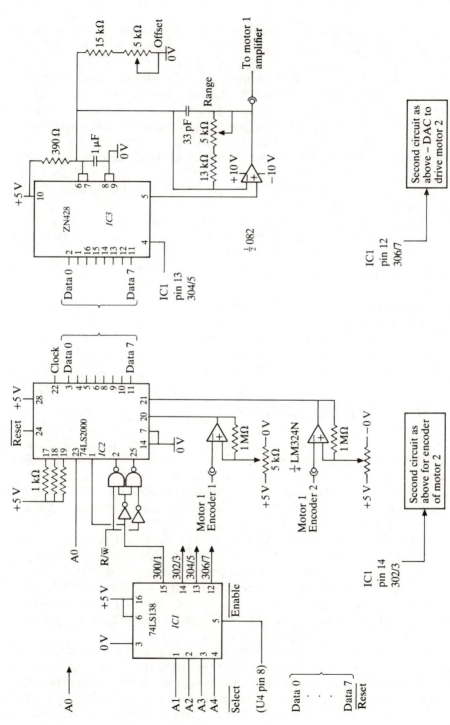

Figure 4-2 Interface diagram.

45

computer's data bus with the outside world. (Unfortunately on the version I have, this logic signal is not brought to a very accessible point. It is found on pin 8 of IC U4.)

This decoding allows any of the lower five address lines to take any value; these address lines must be further decoded to select the variety of devices that we intend to connect. Since the 74LS2000 has two internal registers which we can address, the lowest address bit is connected to both such chips. A 74LS138 three-to-eight decoder now lets us separate eight pairs of addresses by connecting the decoded logic signal mentioned above to the enable line and address bits 1, 2 and 3 to the inputs. The resulting I/O addresses which activate the output lines (low) are thus the hexadecimal pairs 300/1, 302/3, 304/5 and so on.

If we connect the first two such lines to the enable pins of the 74LS2000 and the next two lines to a pair of digital/analogue converters, we will find the necessary functions at the following addresses (see Fig. 4-2):

High byte of shoulder   position: &H300
Low byte of shoulder   position: &H301
High byte of elbow   position: &H302
Low byte of elbow   position: &H303
Shoulder motor velocity demand: &H304
Elbow motor velocity   demand: &H306

4-4 MOTOR SPEED CONTROL

The command to the motor is passed via a digital/analogue converter. If the value 75 is output to address &H304, it will be latched into a chip that will control an output voltage to be a corresponding value. Storing 0 will give the most negative value, while 255 will give the most positive. To set the voltage to mid-scale, we must output 128, and this is conveniently arranged to leave the motor stationary.

It is easy to see that this voltage can in turn control the motor voltage, but how can we say with any conviction that we are controlling motor speed? We need to add some extra velocity signal to achieve a reasonably tight 'speed loop'. For a 'high class' system a tachometer would be added to the motor to generate a clean velocity signal. This can be costly. Here we make use of a much cheaper alternative, which still gives a more than acceptable performance.

As the speed of a permanent magnet d.c. motor increases, so does the generated *back-e.m.f.* If a voltage supply is connected to an unloaded motor, its speed will increase until the generated voltage is just short of the applied voltage; there will be just enough difference to draw sufficient current to overcome the losses in the motor. As the motor becomes mechanically loaded, so its speed will drop to allow more and more current to be drawn. This is speed control of a sort, but extra feedback will much improve its performance.

The voltage across the motor is the sum of its back-e.m.f. and the IR voltage of the drive current through the armature resistance. If a small resistor is added in series with the motor and two large resistors are connected in series between the amplifier output and ground return, we have the bridge circuit of Fig. 4-3. Suitably balanced, it will give no output for variation of the amplifier signal if the motor is held stationary. Release the motor, however, and there will be an output proportional to the back e.m.f. and hence motor velocity.

This signal can be amplified and added into the motor control loop to give a velocity control several times stiffer than the straightforward voltage drive of the motor.

The experiment is arranged to give the student confidence that the computer can output a velocity command signal, to ensure that the position registers can be read and combined into an error signal, and finally to close a control loop in a simple way. No filtering functions are attempted, since these would require a higher realm of 'real time' operation. We hope that the computer will get around to supplying a command signal every few milliseconds or less, but do not depend on these operations being regular. If we aspired to digital dynamic compensation, of the type to be described in Chapter 11, we would need to be sure that the computer dealt with the control at regular, known intervals. That would call for something more ambitious than the simple BASIC program considered here.

To put the experiment into perspective, it is perhaps easiest to reproduce part of the instruction sheet.

4-5 STUDENT NOTES FOR THE SCARA EXPERIMENT

1. Introduction

The two-axis mechanism of the experiment presents the most difficult control aspects of the SCARA robot. It only lacks a vertical movement and rotary gripper to complete a system widely becoming accepted for pick-and-place applications.

The point of the experiment is to investigate the problem of digital position control to achieve fast response and smooth settling to a commanded position. The output positions are tracked by optical transducers, producing pulse trains that are monitored by two special microcircuits. These have a microcomputer-bus compatible output to give position to sixteen-bit precision. The motor drive amplifiers are commanded by digital/analogue converters, and velocity feedback is added to damp the motor responses.

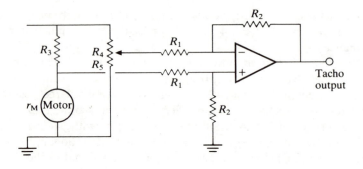

The potentiometer R_4/R_5 is adjusted to make

$$\frac{R_5}{R_4} = \frac{r_M}{R_3}$$

$$\text{Output} = \frac{R_2}{R_1}\,\frac{R_3}{R_3 + r_M} \times \text{motor back-e.m.f.}$$

(a)

Figure 4-3 Drive amplifier showing pseudo-tacho feedback. (a) Principle of pseudo-tacho feedback. (b) Complete motor drive circuit.

That leaves a gap to close between the measured position and the digitally demanded motor velocity. Therein lies the experiment.

An IBM personal computer is used as the controlling microcomputer, although a dedicated system would be used in a practical industrial robot. The IBM allows one-line commands to be used to read the count registers, to set drive levels and even to apply dynamic test signals.

2. Interface details

The controller is interfaced through a 'prototype card'. This provided eight bits of data bus, five bits of address bus, read, write and clock lines, and a select line which responds to any address &H300 to &H31F. A decoder selects the individual I/O devices as follows:

(a) The two digital/analogue converters are found at addresses &H304 and &H306. &H304 drives the 'shoulder' motor, while &H306 drives the elbow. Since the eight-bit output values cover the range 0 to 255 and must correspond to both positive and negative drives, the zero drive value is 128. Thus the direct command:

OUT(&H304,128) : OUT(&H306,128)

will leave both motors at rest. If now you enter

OUT(&H304,0)

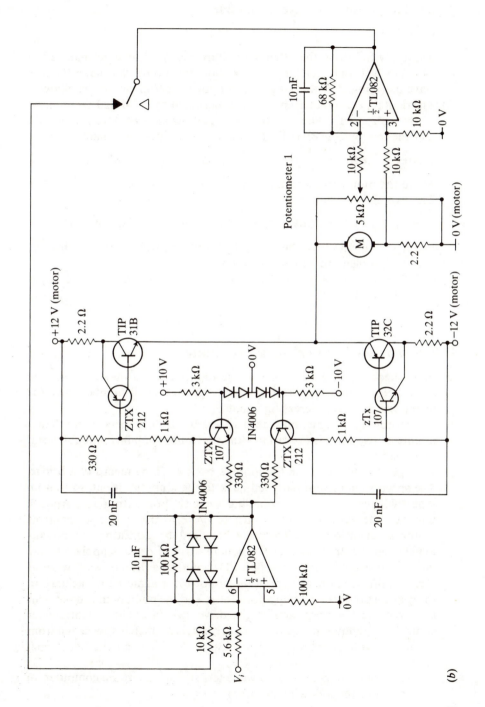

Potentiometer 1

$\frac{1}{2}$ TL082

$\frac{1}{2}$ TL082

(b)

49

the elbow motor will drive hard, while

OUT(&H304,255)

will apply full drive the other way. Start gently with values like 110 or 145. You will find that driving the shoulder motor can make the arm wave alarmingly; at this stage restrict your experiments to the elbow.

(b) The digital position of each motor is measured by a 74LS2000 chip from Texas Instruments. The high byte of the elbow motor is found at &H302 and the low byte at &H303. Try the one-line direct command:

WHILE 1 : PRINT INP(&H302),INP(&H303) : WEND

Move the motor by hand and see the numbers change.
The two eight-bit numbers can be combined as follows:

WHILE 1 : PRINT 256∗INP(&H302) + INP(&H303) : WEND

and if you again move the motor by hand you will see the result—and another problem, to be dealt with later.

The position of the shoulder motor is given by &H300 (high byte) and &H301 (low byte).

3. Procedure for first single-axis experiment

(a) First check out the operation of the inputs and outputs as described above. Note the drive directions for large and small numbers and the directions that give increasing counts.

(b) The motor speed signal may need setting up. *Do not* spend more than a minute or two on this operation—the previous group has probably set it up already.

Speed is derived from the motor's back-e.m.f. by means of a bridge. The speed signal is normally fed back to the amplifier input, so that an input will result in a corresponding velocity. For setting-up purposes, however, the feedback switch must be *open*, so that the input controls motor acceleration. The bridge is balanced by adjusting the potentiometer on the driver board. If a square wave of drive is applied to the motor, the velocity signal (test point 1) should be a triangle wave without steps. In practice, the rattling of the motor rotor against the backlash in the gearing produces a 'lumpy' waveform, and the point of the adjustment is to remove the crisp step as the drive switches. The potentiometer controls the amount of square wave drive signal which is subtracted from the motor voltage, removing the i.r. term and leaving the back-e.m.f. alone.

How can you get a square wave test signal? From the computer, of course. Enter the following program:

```
10 DAC = &H304 : V = 40 : T = 200
20 WHILE 1
30 OUT(DAC,128 + 4) : V = − V
40 FOR I = 0 TO T : NEXT I
50 WEND
```

Type RUN and the motor should start to vibrate. If the amplitude or speed are unsuitable, break the program and edit line 10 to change the values of V or T. If the motor position drifts, adjust the 'offset' potentiometer on the interface board. Monitor test point 1 on an oscilloscope and adjust potentiometer 1 until there are no sharp steps in the speed signal.

Set the drive to 128 (neutral) and move the forearm gently by hand. Note the change in resistance to motion when the feedback switch is opened and closed. After setting the feedback with loving care, reset it the pragmatic way: while still moving the forearm by hand with the feedback switch closed, adjust the feedback potentiometer either way. One way the arm will become less stiff; the other way a point will be reached at which the arm will become unstable. Find this point, reduce the feedback 'enough for a safety margin', check for 'reasonable' stiffness and continue with the experiment.

(c) Now the position signal must be read in a usable form. One way to combine the two bytes is by letting $X = 256*INP(\&H302) + INP(\&FCC3)$. You should already have tried this in following the directions of 2(b). You will notice that the position will read zero at the angle where power was first applied to the system—the point at which the counter was reset. As the angle is moved one way, the count increases steadily. In the other direction, however, the count first jumps to 65535 before reducing. This discontinuity will cause control problems and must be corrected.

An effective way of setting the reset point to the middle of the linear range is to exclusive-or the high byte with 128. Instead of jumping from zero to 255, the high byte value will change from 128 to 127. Try out:

```
WHILE 1 : PRINT 256*(INP(&H302) XOR 128) + INP(&H303) :
                                                    WEND
```

There is another potential problem arising from the division of the count into two bytes. Suppose the counts on the two addresses are decreasing as follows:

1	1
1	0
0	255
0	254

If the high byte is read just before it changes value and the low byte is read afterwards, then the result will read

1   255

giving a reading error of 256 counts. Reversing the order of reading will only make an upward count dangerous instead.

The solution is to read the count using machine code (for speed), to read the high byte and then the low byte, and then to read the high byte a second time to ensure that it has not changed in the meantime. If it has changed, start reading afresh. We will try the 'fix' later, after seeing the 'glitch'.

Begin investigating the position control problem by applying proportional control. The position can be read by a subroutine as follows:

```
10 HI = &H302 : LO = &H303 : DAC = &H306 : Z7 = 128 :
                                                Z8 = 256
1000 X = Z8*(INP(HI) XOR Z7) + INP(LO) : RETURN
```

To apply proportional control you will need to output a speed command proportional to the error, $Y-X$, where Y is the target point. You only have a range of ± 127 in the output before it limits, so use a small value of GAIN to start with. The program first sets the target to be the arm's present position:

```
20 GAIN = 0.05
30 GOSUB 1000 : Y = X

100 WHILE 1 : GOSUB 1000 : OUT(DAC,Z7 + GAIN*(Y − X) :
                                                WEND
```

Enter the five lines of the program and type RUN. If you move the arm by hand, the motor will drive it back to its original position. Do not use excessive force, but estimate the 'stiffness' of the control for various values of GAIN—edit line 20.

Now try the response of the system to a step of error. Run the program and break it. Move the arm slightly and then type

GOTO 100

(following it with ⟨return⟩, of course). You will find that for values of gain around unity, any substantial initial deflection will give an error message. You will have exceeded the drive limit. The next refinement of the program is to include a limit in the software. The following few lines will convert a signed drive value into the range required for the DAC, limiting it to 0 or 255 if the range is exceeded:

```
1010 SIGNAL = DRIVE + Z7
1020 IF SIGNAL > &FF THEN SIGNAL = &FF
1030 IF SIGNAL < 0 THEN SIGNAL = 0
1040 OUT(DAC,SIGNAL) : RETURN
```

Now line 100 is modified to take advantage of it:

```
100 WHILE 1 : GOSUB 1000 : DRIVE = GAIN*(Y − X) :
                                    GOSUB1010 : WEND
```

You can then choose a higher value of GAIN and achieve a stiffer system. For high GAIN values, however, you will find that the step response will result in overshoots or will become unstable.

(d) You can exploit the presence of the computer to obtain a plot of the step response. The following program indicates an outline. Note that lines 100 to 120 are doing all the controlling, with the aid of the functions from 1000 on:

```
10 HI = &H300 : LO = &H301 : DAC = &H304 : Z7 = 128 :
                                    Z8 = 256 : Z9 = 512
20 OUT(DAC.Z7) : NP = 200 : SCX = 320/NP
30 DIM EE(NP) : REM ARRAY TO REMEMBER RESPONSE
80 INPUT "STEP, GAIN" Y,GAIN : SCY = 100/ABS(Y)
90 GOSUB 1000 : Y = Y + X : IF SCY > 1 THEN SCY = 1

100 FOR I = 0 TO NP
110 GOSUB 1000 : E = Y − X : DRIVE = (GAIN*E) :
                                    GOSUB1010 : EE(I) = E
120 NEXT I

200 SCREEN 1
210 LINE (310,100) − (0,100)
220 FOR I = 0 TO NP
230 LINE − (SCX*I,SCY*(1 − EE(I)))
240 NEXT I
250 PRINT " STEP ="; ABS(EE(0)),"GAIN = "; GAIN : GOTO 80

1000 X = Z8*(INP(HI) XOR Z7) + INP(LO) : RETURN

1010 SIGNAL = DRIVE + Z7
1020 IF SIGNAL > &FF THEN SIGNAL = &FF
1030 IF SIGNAL < 0 THEN SIGNAL = 0
1040 OUT(DAC,SIGNAL) : RETURN
```

Make comparisons of the responses for different gains and step sizes, choosing a 'best' gain for each order of step size.

(e) You will certainly see some of the 'glitches' described above in the form of spikes on the response for either a positive or a negative step. The 'fix' is to redefine the subroutine which reads the encoder as follows:

```
1000 Z = −1 : WHILE Z<>INP(HI) : Z = INP(HI) :
                                  ZZ = INP(LO) : WEND
1005 X = Z8*(Z EOR Z7) + ZZ : RETURN
```

Now the software keeps rereading the encoder until it sees the same value for the high byte both before and after reading the low byte. Try it again with the new function.

(*f*) Now you can transfer your attention to the shoulder joint simply by changing the addresses in line 10 to become

```
10 HI = &H302 : LO = &H303 : DAC = &H306 : Z7 = 128 :
                                  Z8 = 256 : Z9 = 512
```

You should find that because of the differing moment of inertia, a different value of gain gives the best result.

(*g*) Finally it is time to drive both axes together, noting their dynamic interaction. Instead of saving the responses for plotting as a continuous line, the results can be displayed on the screen at once in dot form, using the PSET command. The change to the program is not great:

```
10 EHI = &H300 : ELO = &H301 : EDAC = &H304 : Z7 = 128 :
                                  Z8 = 256 : Z9 = 512
20 SHI = &H302 : SLO = &H303 : SDAC = &H306
30 OUT(SDAC,Z7) : OUT(EDAC,Z7) : NP = 320 :
                                  SCX = 320/NP
40 INPUT "    ELBOW STEP, GAIN" EY,EGAIN
50 INPUT "SHOULDER STEP, GAIN" SY,SGAIN
60 IF ABS(EY)>ABS(SY) THEN SCY = 100/ABS(EY) ELSE
                                  SCY = 100/ABS(EY)
70 IF SCY>1 THEN SCY = 1
80 GOSUB 1000 : EY = EY + EX : GOSUB2000 : SY = SY + SX
90 SCREEN 1 : LINE (310,100) − (0,100)

100 FOR I = 0 TO NP
110 GOSUB 1000 : EE = EY − EX : EDRIVE = (EGAIN*EE) :
                                  GOSUB1010
120 PSET(SCX*I,SCY*(1 + EE))
130 GOSUB 2000 : SE = SY − SX : SDRIVE = (SGAIN*SE) :
                                  GOSUB2010
140 PSET(SCX*I,SCY*(1 + SE))
150 NEXT I
160 END

1000 Z = −1 : WHILE Z<>INP(EHI) : Z = INP(EHI) :
                                  ZZ = INP(ELO) : WEND
1005 EX = Z8*(Z EOR Z7) +ZZ : RETURN
```

```
1010 SIGNAL = EDRIVE + Z7
1020 IF SIGNAL>&FF THEN SIGNAL = &FF
1030 IF SIGNAL<0 THEN SIGNAL = 0
1040 OUT(EDAC,SIGNAL) : RETURN

2000 Z = -1 : WHILE Z<>INP(SHI) : Z = INP(SHI) :
                                  ZZ = INP(SLO) : WEND
2005 SX = Z8*(Z EOR Z7) +ZZ : RETURN

2010 SIGNAL = SDRIVE + Z7
2020 IF SIGNAL>&FF THEN SIGNAL = &FF
2030 IF SIGNAL0 THEN SIGNAL = 0
2040 OUT(SDAC,SIGNAL) : RETURN
```

4-6 CLOSING REMARKS

On the BBC computer used originally, it was easy to include sections of machine code for 'debouncing' the encoder signals and for implementing the limiters. On the IBM this is more complicated and might not be thought to be worth the effort entailed. If speed of running is seen as a limitation, then the BASIC program can be compiled. It must be admitted, though, that the greater benefit lies in giving confidence to the student that a feedback loop can be closed with but a few bytes of code.

This material has been included more to indicate how easily digital feedback experiments can be set up than as a recipe book for a laboratory course. However, further details of the circuitry and software can be obtained from the author at Portsmouth Polytechnic, Anglesea Road, Portsmouth PO1 3DJ, UK.

FREQUENCIES, REAL AND COMPLEX

P5-1 INTRODUCTION

Sine waves and allied signals are the tools of the control trade. They can be represented in terms of their amplitude, frequency and phase. After a page or two of algebraic trigonometry extracting phase angles from mixtures of sines and cosines, however, it becomes clear that some computational short-cuts are more than welcome. This section is concerned with the representation of sine waves as imaginary exponentials, and the extention to include the interpretation of complex exponentials.

Notation   Frequency terms are represented by the angular frequency omega, where $\omega = 2\pi f$. The real part of a complex frequency is σ (sigma).

P5-2 CONSEQUENCES OF THE DEMOIVRE THEOREM

$$\cos(\omega t) = [\exp(j\omega t) + \exp(-j\omega t)]/2$$
$$\sin(\omega t) = [\exp(j\omega t) - \exp(-j\omega t)]/2j \qquad \text{(P5-1)}$$

The statements above make a bold start to the section, but perhaps some effort should be made to prove them. One approach, common in school textbooks, is to derive the power series for $\sin(x)$ and $\cos(x)$, and also for $\exp(x)$. The coefficient of x^n in $\exp(x)$ is of course $1/n!$. Careful scrutiny of the sine and cosine series show that alternate terms are present, with alternating signs:

Function			Coefficients of x^n				
$\exp(x)$	1	1/1!	1/2!	1/3!	1/4!	1/5!	...
$\cos(x)$	1	0	$-1/2!$	0	1/4!	0	...
$\sin(x)$	0	1/1!	0	$-1/3!$	0	1/5!	...
$\exp(jx)$	1	$j/1!$	$-1/2!$	$-j/3!$	1/4!	$j/5!$	...

Multiply the sine series by j, add the cosine term by term and inspection shows that

$$\cos(x) + j\sin(x) = \exp(jx)$$

that is

$$\cos(\omega t) + j\sin(\omega t) = \exp(j\omega t) \tag{P5-2}$$

Subtracting the series also shows that

$$\cos(\omega t) - j\sin(\omega t) = \exp(-j\omega t)$$

Not much mathematical effort is needed to derive the results of equations (P5-1) from these equations.

There is an alternative proof (or demonstration) much closer to the heart of a control engineer. It depends on the techniques for solving linear differential equations touched upon in the prelude to Chapter 2. It depends also on the concept of the uniqueness of a solution satisfying enough initial conditions.

Let $y = \cos(\omega t)$. The value of y at $t = 0$ is 1. If we differentiate with respect to t, we have

$$\dot{y} = -\omega \sin(\omega t)$$

and so y has zero value at $t = 0$. Differentiating again gives

$$\ddot{y} = -\omega^2 \cos(\omega t)$$

from which we see that

$$\ddot{y} = -\omega^2 y$$

Now, faced with the second-order differential equation

$$\ddot{y} + \omega^2 y = 0$$

we try out the solution $y = \exp(mt)$ and arrive at

$$m^2 y + \omega^2 y = 0$$

from which we deduce that

$$m = \pm j\omega$$

We have a general solution

$$y = A \exp(j\omega t) + B \exp(-j\omega t)$$

and we must choose A and B to satisfy the initial conditions $y = 1$ and $\dot{y} = 0$ at $t = 0$. Since the exponential terms each have the value 1 at $t = 0$ while their derivatives have values $j\omega$ and $-j\omega$ respectively, $A = B = \frac{1}{2}$. The first result is proved. The second equation follows in the same way, except that the initial value of $\sin(\omega t)$ is zero, while its derivative at $t = 0$ is ω.

P5-3 COMPLEX AMPLITUDES

The equations of (P5-1) allow us to exchange the ungainly sin and cos functions for more manageable exponentials, but we are still faced with terms in both $+j\omega t$ and $-j\omega t$. Can this be simplified?

Equation (P5-2) shows $\exp(j\omega t)$ as a mixture of the sine and cosine functions. The real part, however, is the cosine alone. If we multiply $\exp(j\omega t)$ by $-j$, we find a real part which is now the sine. If we multiply it by a complex number, we can clearly obtain the sum of sine and cosine in any proportion we wish; in other words we can handle sine waves at a variety of phase angles. On the strict understanding that we take the real part of any expression when describing a function of time, we can now deal in complex amplitudes of $\exp(j\omega t)$.

Algebra of addition and subtraction will clearly work out without any complications. The real part of the sum of $(a + jb) \exp(j\omega t)$ and $(c + jd)$ $\exp(j\omega t)$ is seen to be the sum of the individual real parts, i.e. we can add the complex numbers $(a + jb)$ and $(c + jd)$ to represent the new mixture of sines and cosines. Beware, however, of multiplying the complex numbers to denote the product of two sine waves. For anything of that sort you must go back to the precise representation given in equations (P5-1).

Another operation which is linear, and therefore in harmony with this representation, is differentiation. It is not hard to show that

$$\frac{d}{dt}(a + jb)\exp(j\omega t) = j\omega(a + jb)\exp(j\omega t)$$

Differentiation a second time is still a linear operation, and so each time the mixture is differentiated we obtain an extra factor of $j\omega$.

The simpiifying effect that this has on solving differential equations for steady solutions with sinusoidal forcing functions is enormous. In the 'knife and fork' approach we would have to assume a result of the form $A \cos(\omega t) + B \sin(\omega t)$, substitute this into the equations and unscramble the resulting mess of sines and cosines. Let us try an exercise to see the improvement.

Exercise P5-3-1   The system described by the second-order differential equation

$$\ddot{x} + 4\dot{x} + 5x = u \qquad\qquad (P5\text{-}3)$$

is forced by the function

$$u = \sin(3t) + 2\cos(3t)$$

What is the steady state solution (after the effect of any initial transients has died away)?

The solution will be a mixture of sines and cosines of $3t$, which we can represent as the real part of $X \exp(3jt)$. The derivative of x will be the real part of $3jX \exp(3jt)$, while the second derivative will be the real part of $(-9)X \exp(3jt)$. At the same time, u can be represented as the real part of $(2 - j) \exp(3jt)$. Substituting all these into Eq. (P5-3) produces a litter of multiples of $\exp(3jt)$, and so we can take the exponential factor out of the equation and just equate the coefficients. We get

$$(-9)X + 4(3j)X + 5X = (2 - j)$$

that is

$$(-4 + 12j)X = (2 - j)$$

and so

$$X = \frac{2 - j}{-4 + 12j}$$

$$= \frac{(2 - j)(-4 - 12j)}{160}$$

$$= \frac{-20 - 20j}{160}$$

$$= \frac{-1 - j}{8}$$

So we see that the final solution is

$$x = -\tfrac{1}{8}\cos(3t) + \tfrac{1}{8}\sin(3t)$$

As a masochistic exercise, solve the equation again the hard way, without using complex notation.

P5-4 MORE COMPLEX STILL

We have tried out the use of complex numbers on the amplitudes of sinusoids. Could we usefully consider complex frequencies too? Yes, anything goes.

How can we interpret $\exp[(\sigma + j\omega)t]$? It expands at once to give $\exp(\sigma t)$ $\exp(j\omega t)$; in other words the imaginary exponential which we now know as a sinusoid is multiplied by a real exponential of time. If the value of σ is negative, then the envelope of the sinusoid will decay towards zero, after the fashion of the clang of a bell. If σ is positive, however, the amplitude will swell indefinitely.

The process of differentiation is still linear, and we see that

$$\frac{d}{dt}(a + jb)\exp[(\sigma + j\omega)t] = (a + jb)(\sigma + j\omega)\exp[(\sigma + j\omega)t]$$

In other words, we can consider forcing functions that are products of sinusoids and exponentials, and can take exactly the same algebraic short-cuts as before (see Fig. P5-1).

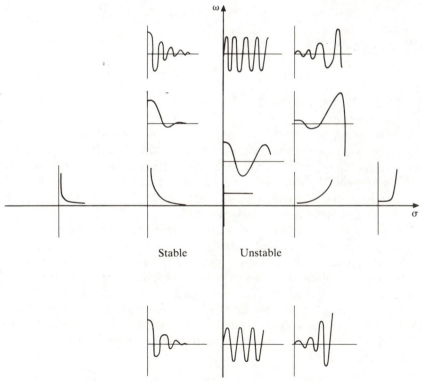

Figure P5-1 Complex frequencies—waveforms corresponding to points in the s-plane. In the left half-plane, transients decay to zero. In the right half-plane, they become infinite.

Exercise P5-4-1 *Consider Exercise P5-3-1 above when the forcing function is instead $\exp(-t)\cos(2t)$.*

Exercise P5-4-2 *Consider Exercise P5-3-1 again with an input function $\exp(-2t)\sin(t)$. What went wrong? Read the chapter to find out.*

FIVE

TESTING THE UNKNOWN

5-1 INTRODUCTION

It was tempting to call this chapter 'whistling in the dark'. The control engineer does not always know the intimate details of the dynamic equations of the system to be controlled, and must sometimes treat it as a 'black box' with input and output but with unknown contents. To explore the characteristics of the system, it must be provoked with a test input signal, and this was traditionally a sinusoidal 'whistle'.

In the early sixties, an aircraft autopilot relied heavily on electromechanical components for its computation, using synchro resolvers to perform coordinate transformations. Signals were carried as modulations of a 400 Hz carrier, and the piece of test equipment to produce such a signal with low frequency sinusoidal data also contained servomotors and a synchro resolver. It was affectionately known as a 'whistle-box', not least for the 400 Hz whine of the servos. What is so attractive about a sinusoidal test signal when a step change in amplitude of the 400 Hz carrier was available at the flick of a switch?

5-2 EIGENFUNCTIONS AND GAIN

In the main, control theory is concerned with linear systems. That is to say, the differential equations contain constants, variables and their derivatives of various orders, but never the products of variables—be they states or inputs.

The derivative of a sine wave is another sine wave (cosine wave) of the same frequency, shifted in phase and probably changed in amplitude. The sum of two sine waves of the same frequency and assorted phases will be yet another sine wave of the same frequency, with phase and amplitude which can be found by a little algebra. However many derivatives we add, if the basic signal is sinusoidal then the mixture will also be sinusoidal.

If we apply a sinusoidal input to a linear system, allowing time for transients to die away, the output will settle down to a similar sinusoid. If we double the size of the input, the output will also settle to double its amplitude, while the phase relationship between input and output will remain the same. The passage of the sine wave through the system will be characterized by a 'gain', the ratio between output and input, and a phase shift.

In the prelude to this chapter, we saw that a phase shift can be represented by means of a complex value of gain when the sine wave is expressed in its complex exponential form $\exp(j\omega t)$.

Now we see that applying such a signal to the input of the linear system will produce exactly the same sort of output, multiplied only by a constant gain (admittedly complex). The signal $\exp(j\omega t)$ is an *eigenfunction* of any linear system.

Altering the frequency of the input signal will of course change the gain of the response; the gain is a function of the test frequency. If tests are made at a variety of frequencies, then a curve can be plotted of gain against frequency, the *frequency response*. As we will see, a simple series of such tests can in most cases give all the information needed to design a stable controller.

The sine wave is not alone in the role of eigenfunction. Any exponential $\exp(st)$ will have similar properties: that for an input of $\exp(st)$ and for the correct initial conditions the system can give an output $G(s)\exp(st)$. Now $G(s)$ is the gain for that particular value of the constant s. Clearly if s is real, $\exp(st)$ is a less convenient test signal to use than a sine wave. If s is positive, the signal will grow to enormous proportions. If s is negative, the experimenter will have to be swift to catch the response before it dies away to nothing. Although of little use in experimental terms, the mathematical significance of these more general signals is very important, especially when s is allowed to be complex.

5-3 A SURFEIT OF FEEDBACK

In the case of a position servo, it is natural to seek as high a 'loop gain' as possible; that is to say, the correction motor should exert a large torque for as small an error as possible. When the dynamics are accurately known in state-space form, feedback can be determined by analytical considerations. In the absence of such insight, early engineers had to devise practical methods of finding the limit to which simple feedback gain could be increased. These are, of course, still of great practical use today.

Feedback has many more roles in electronics, particularly for reducing the non-linearity of amplifiers and for reducing uncertainty in their gains. An amplifier stage might have a gain, say, with a value between 50 and 200. Two such stages could give a range of combined gains between 2500 and 40 000—a factor of 16. Suppose we have an accurate target gain of 100; how closely can we control it?

We apply feedback by mixing a proportion k of the output signal with the input. To work out the resulting gain, it is easiest to work backwards. To obtain an output of 1 volt, the input to the amplifier must be $1/G$ volts, where G is the gain of the open loop amplifier. We now feed back a further proportion k of the output, in such a sense as to make the necessary input greater (negative feedback).

Now we have

$$V_{in} = \left(\frac{1}{G} + k \right) V_{out}$$

which can be rearranged to give the gain, the ratio of output to input, as

$$\frac{V_{out}}{V_{in}} = \frac{G}{1 + kG}$$

To check on the accuracy of the result, this can be rearranged as

$$\frac{1}{k} \frac{1}{1 + 1/(kG)}$$

If we are looking for a gain of 100, $k = 0.01$. If G lies between 2500 and 40 000, then kG is somewhere between 25 and 400. We see that the closed loop gain may be just four per cent low if G takes the lowest value; the uncertainty in gain has been reduced dramatically. The larger the minimum value of kG, the smaller will be the uncertainty in gain, and so a very large loop gain would seem to be desirable.

Of course, positive feedback will have a very different effect. The feedback now assists the input, increasing the closed loop gain to a value

$$\frac{G}{1 - kG}$$

If k starts at a very small value and is progressively increased, something dramatic happens when $kG = 1$. Here the closed loop gain becomes infinite; the output can flip from one extreme to the other at the slightest provocation. If k is increased further, the system becomes *bistable*, giving an output at one extreme limit until an input opposes the feedback sufficiently to flip it to the other extreme.

All would be well with huge loop gains if the response of the amplifier were infinitely fast, but unfortunately it will contain some dynamics. The open loop gain is not a constant G, but is seen to be a function of the applied test frequency $G(j\omega)$ complete with phase shift. In any but the simplest model of the amplifier, this phase shift can approach or reach 180°, and that is where trouble can break out. A phase shift of 180° is equivalent to a reversal in sign of the original sine wave. Negative feedback becomes positive, and if the value of kG still has magnitude greater than unity, then closing the loop will certainly result in oscillation.

The determination of a permissible level of k will depend on the race between increasing phase shift and diminishing gain as the test frequency is increased. We could measure the phase shift at the frequency where kA just falls below unity. As long as this is below 180° we have some margin of safety—the actual shortfall being called the *phase margin*. Alternatively, we could measure the gain at the frequency that gives just 180° phase shift. The amount by which kG falls below unity here we can term the *gain margin*. These led first to rules of thumb and then via an art to a science. Now we can put the methods onto a firm foundation of mathematics.

5-4 POLES AND POLYNOMIALS

Analysis of the servomotor problem from the state-space point of view gave us a list of first-order differential equations. A *lumped linear system* will similarly have a set of state equations, each having a simple d/dt on the left and a linear combination of variables on the right. There are of course many other systems that fall outside such a description, but why look for trouble.

$$\dot{\mathbf{x}} = A\mathbf{x} + B\mathbf{u}$$

$$y = C\mathbf{x}$$

For a start let us suppose that the system has a single input and a single output. With a certain amount of algebraic juggling we can eliminate all the x's from the equations and get back again to the 'traditional' form of a single higher order equation linking input and output. This will be of the form:

$$\frac{d^n y}{dt^n} + a_1 \frac{d^{n-1} y}{dt^{n-1}} + \cdots + a_n y = b_0 \frac{d^m u}{dt^m} + \cdots + b_m u \qquad (5\text{-}1)$$

Now let us try the system out with an input that is an exponential function of time $\exp(st)$—without committing ourselves to stating whether s is real or complex. If we assume that the initial transients have all died away, then y will also be proportional to the same function of time. Since the derivative of $\exp(st)$ is the same function, multiplied by s, all the time derivatives simply turn into powers of s. We end up with

$$(s^n + a_1 s^{n-1} + \cdots + a_n)y = (b_0 s^m + \cdots + b_m)\exp(st) \qquad (5\text{-}2)$$

The gain, the ratio between output and input, is now the ratio of these two polynomials in s. If we commit ourselves to making s be pure imaginary, with value ($j\omega$), we obtain an expression for the gain (and phase shift) at any frequency.

Any polynomial can be factorized into a product of linear terms of the form

$$(s - p_1)(s - p_2)(s - p_3) \ldots$$

where the coefficients are allowed to be complex. Clearly, if s takes the value p_1, then the value of the polynomial will become zero. What if the polynomial in question is the denominator of the expression for the gain? Does it not mean that the gain at complex frequency p_1 is infinite? I am afraid that it does.

The gain function, in the form of the ratio of two polynomials in s, is more commonly referred to as the *transfer function* of the system, and the values of s that make the denominator zero are termed its *poles*.

It is true that the transfer function becomes infinite when s takes the values of one of the poles, but this can be interpreted in a less dramatic way. The ratio of output to input can just as easily be infinite by defining the input to be zero for a non-zero output. In other words, we can get an output of the form $\exp(p_i t)$ for no input at all, where p_i is any pole of the transfer function.

If the pole has a real, negative value, say -5, it means that there can be an output $\exp(-5t)$. This is a rapidly decaying transient, which might have been provoked by some input before we set the input to zero. This sort of transient is unlikely to cause any problem.

Suppose instead that the pole has value $-1 + j$. The function $\exp[(-1 + j)t]$ can be factorized into $\exp(-t) \exp(jt)$. Clearly it represents the product of a cosine wave of angular frequency unity with a decaying exponential. After an initial 'ping', the response will soon cease to have any appreciable value—all is still well.

Now let us consider a pole that is purely imaginary, $-2j$ say. The response $\exp(-2jt)$ is a continuing sinusoid, never dying away. We are in trouble.

Even worse, consider a pole at $1 + j$. Now we have a sine wave multiplied by an exponential that more than doubles each second. The system is hopelessly unstable.

We conclude that poles that have a negative real part are relatively benign, causing no trouble, but poles that have a real part that is positive or even zero are a sign of instability. What is more, even one such pole among a host of stable ones is enough to make a system unstable.

For now, let us see how this new insight helped the traditional methods of examining a system.

5-5 COMPLEX MANIPULATIONS

The logarithm of a product is the sum of the individual logarithms. If we take the logarithm of the gain of a system described by a ratio of polynomials, we are left adding and subtracting logarithms of expressions no more complicated than $(s - p_i)$, the factors of the numerator or denominator. To be more precise, if

$$G(s) = \frac{(s - z_1)(s - z_2) \cdots (s - z_m)}{(s - p_1)(s - p_2) \cdots (s - p_n)}$$

then

$$\log[G(s)] = \log(s - z_1) + \log(s - z_2) + \cdots + \log(s - z_m)$$
$$- \log(s - p_1) - \log(s - p_2) + \cdots - \log(s - p_n)$$

We are first likely to want to work out a frequency response, by substituting the value $j\omega$ for s, and we are faced with a set of logarithms of complex expressions.

A complex number can be expressed in polar form (Fig. 5-1) as $r \exp(j\theta)$. (Remember that $\exp(j\theta) = \cos\theta + j\sin\theta$.) Here r is the modulus of the number, the square root of the sum of the squares of real and imaginary parts, while θ is the 'argument', an angle in radians whose tangent gives the ratio of imaginary to real parts. When we take the logarithm of this product we see that it splits neatly into a real part, $\log(r)$, and an imaginary part, $j\theta$.

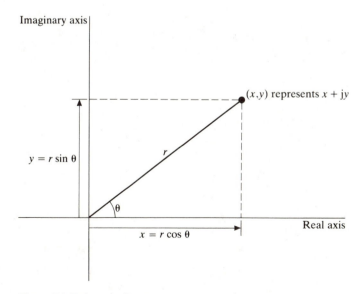

Figure 5-1 Polar coordinates.

Let us consider a system with just one pole, with gain

$$G(s) = \frac{1}{s - p}$$

(Note that for stability, p will have to be negative.) Substitute $j\omega$ for s and we have

$$\log[G(j\omega)] = -\log(j\omega - p)$$

$$= -\log(\sqrt{p^2 + \omega^2}) + j \tan^{-1}\left(\frac{\omega}{p}\right)$$

The real part is concerned with the magnitude of the output, while the imaginary part determines the phase. In the early days of electronic amplifiers, phase was hard to measure. What properties of the system could be deduced from the amplitude alone?

Clearly, for very low frequencies the gain will approximate to $1/(-p)$, i.e. the log gain will be roughly constant. For very high frequencies, the term ω^2 will dominate the expression in the square root, and so the real part of log gain will be approximately

$$-\log(\omega)$$

If we plot log gain against $\log(\omega)$, we will obtain a line of the form $y = -x$, i.e. a slope of -1 passing through the origin. A closer look will show that the horizontal line roughly representing the low frequency gain meets this new

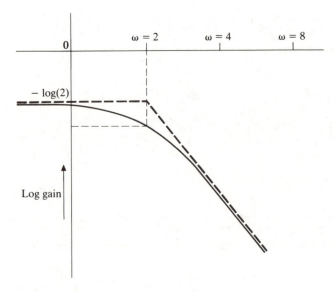

Figure 5-2 Sketched Bode plot of $1/(s + 2)$.

line at the value $\omega = |p|$. These lines, although a fair approximation, do not accurately represent the gain. How far out are they? If we substitute the value p^2 for ω^2 we will have a square root $\sqrt{2}|p|$, i.e. the logarithm giving the real part of the gain will be

$$-\log(|p|) - \log\sqrt{2}$$

The gain at a frequency equal to the magnitude of the pole is thus a factor $\sqrt{2}$ less than the low frequency gain. Take a frequency response; this *breakpoint* will give away the value of the pole. Figure 5-2 is a sketch showing the frequency response when p has the value -2.

5-6 DECIBELS AND OCTAVES

Let us briefly turn aside to some of the traditional terminology you might come across, which could prove confusing.

Remember that the early engineers taking frequency responses were concerned with telephones. They measured the output of the system not by its amplitude but by the power of its sound. This was measured on a logarithmic scale, but the logarithm base was 10. Ten times the output power was one *bel*. A factor of $\sqrt{2}$ in amplitude gives a factor of 2 in power, and is thus $\log_{10}(2)$ bels, or around 0.3 bel. The bel is clearly rather a coarse unit, so we might redefine this as 3 *decibels*. The breakpoint is found when the gain is '3 decibels down'.

We now have gain measured on a logarithmic scale, even if the units are a little strange. Musicians already measure frequency on a logarithmic scale, but the semitone does not really appeal to an engineer as a unit of measurement. Between one 'C' on the piano and the next C higher, the frequency doubles. The unit of log frequency used by the old engineers was therefore the 'octave', a factor of two.

We can plot log power in decibels against log frequency in octaves. What has become of the neat slope of -1 we found above? At high frequencies, the amplitude halves if the frequency is doubled. The power therefore drops by four, giving a fall of 6 decibels per octave. Keep the slogans '3 decibels down' and '6 decibels per octave' safe in your memory!

5-7 FREQUENCY PLOTS AND COMPENSATORS

Let us return to simpler units, and look again at the example of

$$G(s) = \frac{1}{s + 2}$$

We have noted that the low frequency gain is nearly $\frac{1}{2}$, while the high frequency gain is close to $1/\omega$. We have also seen that at $\omega = 2$ the gain has fallen by a factor of $\sqrt{2}$. Note that at this frequency, the real and imaginary parts of the denominator have the same magnitude, and so the phase shift is a lag of 45°—or in radian terms $\pi/4$. As the frequency is increased, the phase lag increases towards 90°.

We can justify our obsession with logarithms by throwing in a second pole, let us say

$$G(s) = \frac{10}{(s + 2)(s + 5)}$$

(see Fig. 5-3). The extra factor of 10 will keep up our low frequency gain to unity. We can consider the logarithm of this gain as the sum of the two logarithms:

$$\log(G) = \log\left(\frac{2}{s + 2}\right) + \log\left(\frac{5}{s + 5}\right)$$

The first is roughly a horizontal line at zero, diving down at a slope of -1 from a breakpoint at $\omega = 2$. The second is similar, but with a breakpoint at $\omega = 5$. Put them together and we have a horizontal line, diving down at $\omega = 2$ with a slope of -1, taking a further nosedive at $\omega = 5$ with a slope of -2. If we add the phase shifts together, the imaginary parts of the logarithmic expressions, we get the following result.

At low frequency, the phase shift is nearly zero. As the frequency reaches 2 radians per second, the phase shift has increased to 45°. As we increase the frequency beyond the first pole, its contribution approaches 90° while the second pole starts to take effect. At $\omega = 5$, the phase shift is around 135°. As frequency increases further, the phase shift approaches 180°. It never *quite* reaches it, so in theory we could never make this system unstable however much feedback we applied. We could have a nasty case of resonance, however, but only at a frequency well above 5 radians per second.

In this simple case, we can see a relationship between the slope of the gain curve and the phase shift. If the slope of the 'tail' is -1, the ultimate phase shift is 90°—no problem. If the slope is -2, we might be troubled by a resonance. If it is -3, the phase shift is heading well beyond 180° and we had better be wary. Watch out for the watershed at 'twelve decibels per octave'.

In this system, there are no phase shift effects that cannot be predicted from the gain. That is not always the case. Veteran engineers lay in dread of *non-minimum-phase* systems.

Consider the system defined by

$$G(j\omega) = \frac{j\omega - 2}{j\omega + 2}$$

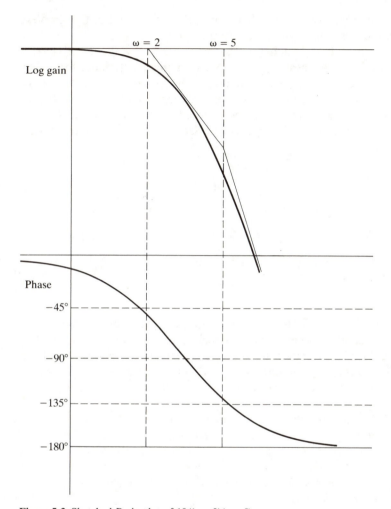

Figure 5-3 Sketched Bode plot of $10/(s + 2)(s + 5)$.

Look at the gain and you will see a magnitude $(\omega^2 + 4)/(\omega^2 + 4)$; the gain is unity at all frequencies. The phase shift is a different matter entirely. At low frequencies there is an inversion—a lead of 180°—with phase lead reducing via 90° at 2 radians per second to no shift at high frequencies. A treacherous system to consider for the application of feedback! This is a classic example of a non-minimum-phase system. (In general, a system is non-minimum phase if it has zeros in the right half of the complex frequency plane.)

The plot of log amplitude against log frequency is referred to as a Bode diagram. It appears that we can add together a kit of breakpoints to predict a response, or alternatively inspect the frequency response to get an insight into the transfer function. In the main this is true, although any but the simplest systems will require considerable skill to interpret.

The Bode plot that includes a second curve showing phase shift is particularly useful for suggesting means of squeezing a little more loop gain or for stabilizing an awkward system. If the feedback is not merely proportional to the output, but instead contains some *compensator* with gain function $F(s)$, then the stability will be dictated by the product of the two gains, $F(s)G(s)$. Following the argument of Sec. 5-3, we see that the closed loop gain will be

$$\frac{G(s)}{1 + F(s)G(s)}$$

and so the poles of the closed loop system are the roots of

$$1 + F(s)G(s) = 0$$

A *phase advance* circuit can be added into the loop of the form

$$F(j\omega) = \frac{3j\omega + a}{j\omega + 3a}$$

This particular compensator will have a low frequency gain of one third and a high frequency gain of three. It might therefore appear to whittle down the gain margin. At second glance, however, it is seen to give a positive phase shift. In this case the phase shift reaches $\tan^{-1}(4/3)$ at frequency a, enabling an awkward second-order system to be tamed without a resonance.

Exercise 5-7-1 *Suppose that the system to be controlled is an undamped motor, appearing as two integrators in cascade. Now $G(s) = 1/s^2$. The phase shift starts off at 180°, so any proportional position feedback will result in oscillation. By adding the phase advance circuit into the loop (Fig. 5-4), a response can be obtained with three equal real roots. As an exercise, work out the algebra to derive their value.*

The technique of the Bode diagram allows us to sketch the likely frequency response given a set of breakpoints. In the age of the computer there is a more certain way of plotting a frequency response to correspond to a list of poles and zeros. A simple program is listed in Fig. 5-5. It runs under GWBASIC.

At its heart is a subroutine, starting at line 200. This evaluates $G(s)$ exactly as described at the start of Sec. 5-5, with the one difference that the numerator can start with the value G0 of the gain factor. Lines 210 and 220 start with the denominator value $(1 + j.0)$ and multiply it successively by each complex value of $(s - p_i)$. If there are any zeros, lines 240 and 250 perform a similar calculation for the numerator, multiplying $(G0 + j.0)$ successively by $(s - z_i)$. The ratio of numerator and denominator is then

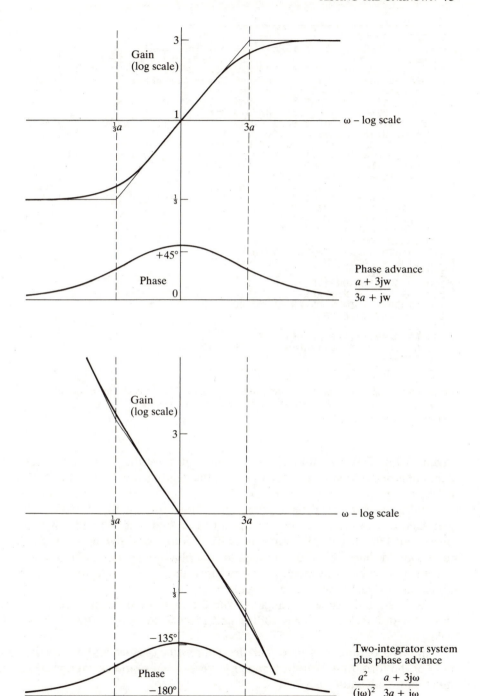

Figure 5-4 Bode plot indicating stabilization of $1/s^2$ using phase advance.

```
  5 SCREEN 3:CLS
 10 NX=50:TDEC=10^(3/NX):REM Frequency increments log scale
 15 REM                       over three decades of freq.
 20 XSC=10:GSC=100:PHSC=80:PH0=120:LG0=120:PI=3.14159:W0=.03
 25 REM Scales for X, gain, phase, Datum for phase, gain
 30 INPUT"how many poles ";NP:DIM P(NP,1)
 40 FOR I=1 TO NP:INPUT"real, imag ";P(I,0),P(I,1):NEXT
 50 INPUT"how many zeroes ";NZ:DIM Z(NZ,1)
 60 FOR I=1 TO NZ:INPUT"real, imag ";Z(I,0),Z(I,1):NEXT
 70 INPUT"Gain factor       ";G0 :CLS
 80 LINE(0,LG0)-(NX*XSC,LG0):REM Unity gain line
 90 LINE(0,PH0+PHSC*PI)-(NX*XSC,PH0+PHSC*PI):REM Phase = -PI
100 LINE(0,LG0)-(0,LG0)
105 W=W0:S=0:FOR X=0 TO NX
110 GOSUB 200:REM Calculate complex gain GR + j GI
120 LG=LOG(GR*GR+GI*GI)/2:REM Log Gain (base e)
130 LINE -(X*XSC,LG0-GSC*LG):W=W*TDEC
140 NEXT:LINE (0,PH0)-(0,PH0)
150 W=W0:S=0:FOR X=0 TO NX
160 GOSUB 200:REM Calculate complex gain GR + j GI
170 PH=ATN(GI/GR):IF GR<0 THEN PH=PH-PI
180 LINE -(X*XSC,PH0-PHSC*PH):W=W*TDEC
190 NEXT:END
200 NR=G0:NI=0:DR=1:DI=0
210 FOR I=1 TO NP:VR=S-P(I,0):VI=W-P(I,1):DR1=DR*VR-DI*VI
220 DI=DR*VI+DI*VR:DR=DR1:NEXT:MD=DR*DR+DI*DI
230 IF NZ=0 THEN 260
240 FOR I=1 TO NZ:VR=S-Z(I,0):VI=W-Z(I,1):NR1=NR*VR-NI*VI
250 NI=NR*VI+NI*VR:NR=NR1:NEXT
260 GR=(NR*DR+NI*DI)/MD:GI=(NI*DR-NR*DI)/MD
270 RETURN
```

Figure 5-5 Listing of Bode plot program. (*Note*: If SCREEN 3 is not available on your machine, edit line 5 to use SCREEN 2 or SCREEN 1. You will then need to adjust the scale factors in line 20.)

taken in line 260, by multiplying the numerator by the conjugate of the denominator and dividing by the square of the denominator's modulus, MD, calculated as an afterthought in line 220.

Poles and zeros are entered at the start of the program, together with the gain factor; then the program plots a unity-gain line (line 80) and a line of phase shift 180°. The 'FOR' loop runs twice, first to enable the log of the gain to be plotted (lines 120, 130) and then for the phase (lines 170, 180). In each case the frequency, W, is increased by multiplying it by a fixed factor, thereby obtaining a logarithmic frequency scale.

As listed, the frequency range is from 0.1 to 100 radians per second—three decades. By editing lines 10, 105 and 150 you can change the range at will.

In this program, the gain calculation subroutine is called with values of s only of the form $(0 + jW)$. In Chapter 7 we will see a similar routine used much more adventurously.

As an exercise, enter and run the program to obtain the various Bode plots of this chapter:

1. One pole: $-2, 0$
   No zeros
   Gain factor 1
2. Two poles: $-2, 0$ and $-5, 0$
   No zeros
   Gain factor 10
3. One pole: $-3, 0$
   One zero: $-0.333, 0$
   Gain factor 3
4. Three poles: $-3, 0$ $0, 0$ and $0, 0$
   One zero: $-0.333, 0$
   Gain factor 3

5-8 SECOND-ORDER RESPONSES

So far we have considered only real poles and zeros. They can also come in complex conjugate pairs, and we had better have a look at the result before moving on. Consider

$$G(s) = \frac{a^2}{s^2 + kas + a^2}$$

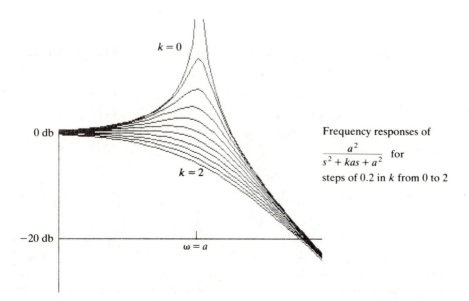

Frequency responses of

$$\frac{a^2}{s^2 + kas + a^2} \text{ for}$$

steps of 0.2 in k from 0 to 2

Figure 5-6 Variation with damping factor of second-order frequency response.

If k has value 2 or greater the denominator will factorize into the product of two terms:

$$(s - p_1)(s - p_2)$$

where p_1 and p_2 are real. If k is less than 2, the roots will form a conjugate complex pair. If we work out the frequency response by substituting $j\omega$ for s, then we see that at $\omega = a$ the phase shift is always 90°. We have

$$G(ja) = \frac{1}{kj}$$

$$= -\frac{1}{k}j$$

If k is steadily reduced, an increasing peak in the response is seen, tending to an infinite amplitude as k tends to zero (see Fig. 5-6).

5-9 EXCITED POLES

An exercise in the prelude to this chapter appeared to require an answer of infinity. A system was excited at a complex frequency corresponding to a pole of the gain function. Should its output really be infinite?

Consider a simpler example, the system described by the gain

$$G(s) = \frac{1}{s + a}$$

when the input is e^{-at}, that is $s = -a$.

If we turn back to the differential equation which the gain function represents, we see that

$$\dot{x} + ax = e^{-at}$$

Since the *complementary function* is now the same as the input function, we must look for a *particular integral* that has an extra factor of t, i.e. the general solution is

$$x = t\,e^{-at} + A\,e^{-at}$$

As t becomes large, we see that the ratio of output to input also becomes large—but the output still tends rapidly to zero.

Even if the system had a pair of poles representing an undamped oscillation, applying the same frequency at the input would only cause the amplitude to ramp upwards at a steady rate; there would be no sudden infinite output. Let one of the poles stray so that its real part becomes positive, however, and there will be an exponential runaway in the amplitude of the output.

MAPPINGS AND TRANSFORMS

P6-1 INTRODUCTION

Complex frequencies and corresponding complex gain functions are now appearing thick and fast. The gains are dressed up as transfer functions, and poles and zeros shout for attention. The time has come to put the mathematics onto a sound footing.

By considering complex gain functions as 'mappings' from complex frequencies, we can exploit some powerful mathematical properties. By the application of Fourier and Laplace transforms to the signals of our system, we can avoid taking unwarranted liberties with the system transfer function, and can find a slick way to deal with initial conditions too.

P6-2 COMPLEX PLANES AND MAPPINGS

A complex number $x + jy$ is most easily visualized as the point with coordinates (x, y) in an *Argand diagram*. In other words, it is a simple point in a plane. The points on the x axis, $y = 0$, are real. The points on the y axis, $x = 0$, are purely imaginary. The rest are a complex mixture.

If we represent $x + jy$ by the single symbol z, then we can start to consider functions of z which will also be complex numbers. For a start, consider

$$w = z^2 \qquad \text{(P6-1)}$$

Here w is another complex number which can be represented as a point in a plane of its own. For any point in the z plane, there is a corresponding point in the w plane. Things get more interesting when we consider the mapping not just of a single point but of a complete line.

If z is a positive imaginary number, ja, then in our example w will be $-a^2$. Any point on the positive imaginary axis of the z plane will map to a point on the negative real axis of the w plane. As a varies from zero to infinity, w moves from zero to minus infinity, covering the entire negative axis. We see that any point on the negative imaginary z axis will also map to the negative real w plane axis. (A mathematician would say that the function mapped the imaginary z axis *onto* the negative real axis, but *into* the entire real axis.)

Points on the z plane real axis map to the w plane positive real axis whether z is positive or negative. To find points that map to the w plane imaginary axis, we have to look towards values of z of the form $a(1 + j)$ or $a(1 - j)$. Other lines in the z plane map to curves in the w plane.

Exercise P6-2-1 *As an exercise, show that* $(1 + ja)$ *maps to a parabola.*

We need to find out more about the way such mappings will distort more general shapes. If we move z slightly, w will be moved too. How are these displacements related? If we write the new value of w as $w + \delta w$, then in Exercise P6-2-1 we have

$$w + \delta w = (z + \delta z)^2$$

that is

$$\delta w = 2z\, \delta z + \delta z^2$$

If δz is small, the second term can be ignored. Around any given value of z, δw is the product of δz and a complex number—in this case $2z$. More generally, we can find a local derivative dw/dz by the same algebra of small increments used in the introduction to differentiation (see Sec. P2-1).

Now when we multiply two complex numbers we multiply their amplitudes and add their 'arguments'. Multiplying any complex number by $(1 + j)$, for example, will increase its amplitude by a factor of $\sqrt{2}$ while rotating its vector anticlockwise by $450°$. Take four small δz's, which take z around a small square in the z plane, and we see that w will move around another square in the w plane, with a size and orientation dictated by the local derivative dw/dz. If z moves around its square in a clockwise direction, w will also move round in a clockwise direction.

Of course, as larger squares are considered, so the w shapes will start to curve. If we rule up the z plane as squared graph paper, the image in the w plane will be made up of *curly squares* (Fig. P6-1).

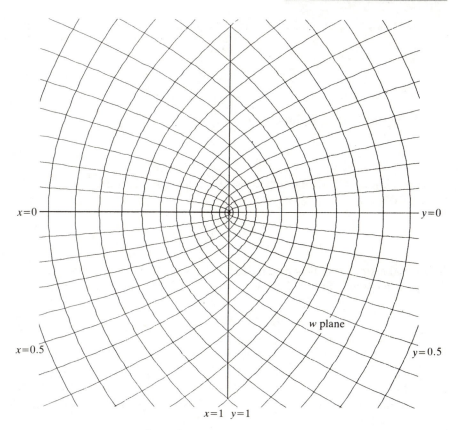

Figure P6-1 Mapping $w = z^2$ as 'curly squares'.

P6-3 THE CAUCHY–RIEMANN EQUATIONS

Above, we suggested that a function of z could be differentiated in the same way that a real derivative could be found. This is true as far as 'ordinary' functions are concerned, ones which would be found in 'real calculus'. However, there are many more functions that can be applied to a complex variable which do not make sense for a real one.

Consider the function $w = \text{Real}(z)$, just the real part, x. Now w will always lie on the real axis, and any thought of it moving in squares is nonsense. This particular function, and many like it, possess no derivative.

If a function $w(z)$ is to have a derivative at any given point, then for a perturbation δz in z we must be able to find a complex number $(A + jB)$ so that around that point $\delta w = (A + jB)\,\delta z$. Let us consider the implications of this requirement, first looking at Exercise P6-2-1.

As w can be expressed as $u + jv$, we can expand the expression to show

$$u + jv = (x + jy)^2$$
$$= (x + jy)(x + jy)$$
$$= x^2 - y^2 + 2jxy$$

Separating real and imaginary parts, we see

$$u = x^2 - y^2$$
$$v = 2xy$$

(P6-2)

In general, u and v will be functions of x and y, a relationship that can be expressed as

$$u = u(x, y)$$

$$v = v(x, y)$$

If we make small changes δx and δy in the values of x and y, the resulting changes in u and v will be defined by the *partial derivatives* in the equations:

$$\delta u = \frac{\partial u}{\partial x} \, \partial x + \frac{\partial u}{\partial y} \, \partial y$$

$$\delta v = \frac{\partial v}{\partial x} \, \delta x + \frac{\partial v}{\partial y} \, \delta y$$

(P6-3)

By holding the change in y to zero, we can clearly see that $\partial u/\partial x$ is the gradient of u if δx is changed alone, and similarly for the other partial derivatives. How

$$\delta w = \delta u + j \, \delta v$$

$$= \frac{\partial u}{\partial x} \, \delta x + \frac{\partial u}{\partial y} \, \delta y + j \left(\frac{\partial v}{\partial x} \, \delta x + \frac{\partial v}{\partial y} \, \delta y \right)$$

$$= \delta x \left(\frac{\partial u}{\partial x} + j \frac{\partial v}{\partial x} \right) + j \, \delta y \left(\frac{\partial v}{\partial y} - j \frac{\partial u}{\partial y} \right)$$

(P6-4)

If w is to have a derivative, we must be able to write

$$\delta w = (\delta x + j \, \delta y)(A + j \, B)$$

$$= \delta x(A + jB) + j \, \delta y(A + jB)$$

so we can equate coefficients with (P6-4) to obtain

$$A = \frac{\partial u}{\partial x} = \frac{\partial v}{\partial y}$$

and
$$B = \frac{\partial v}{\partial x} = -\frac{\partial u}{\partial y}$$

The Cauchy–Riemann equations are

$$\frac{\partial u}{\partial x} = \frac{\partial v}{\partial y} \qquad (P6\text{-}5)$$

and

$$\frac{\partial v}{\partial x} = -\frac{\partial u}{\partial y}$$

The main interest of such mappings from the control point of view is the 'curly squares' approximation for estimating fine detail. The Cauchy–Riemann equations partly express the condition for a function to be 'analytic'. Beware: analytic behaviour may be restricted to portions of the plane, and strange things happen at a singularity such as a pole.

Exercise P6-3-1 *Show that the Cauchy–Riemann equations hold for the functions of the example worked out in Eq. (P6-2).*

P6-4 COMPLEX INTEGRATION

In real calculus, integration can be regarded as the inverse of differentiation. If we want to evaluate the integral of $f(x)$ from 1 to 5, we look for a function $F(x)$ of which $f(x)$ is the derivative. We work out $F(5)-F(1)$, and are satisfied to write this down as the answer. Will the same technique work for complex integration?

In the real case, this integral is the limit of the sum of small contributions

$$f(x)\,\delta x$$

where the total of the δx values takes x from 1 to 5. There is clearly just one answer (for a 'well-behaved' function). In the complex case, we can define a similar integral as the sum of contributions

$$f(x)\,\delta z$$

This time the answer is not so cut and dried. The trail of infinitesimal δz's must take us from $z = 1$ to $z = 5$, but now we are not constrained to the real axis but can wander around the z plane. We have to define the path, or *contour*, along which we intend to integrate.

With luck, Cauchy's theorem can come to our rescue. If we can find a simple curve which encloses a region of the z plane in which the function is everywhere 'regular' (analytic without singularities), then it can be shown that any path between the endpoints lying completely within the region will give the same answer (Fig. P6-2). This is the same as saying that the integral around a loop in the region, starting and ending at the same point, will be zero.

If our integration contour encloses a simple pole we have quite a different

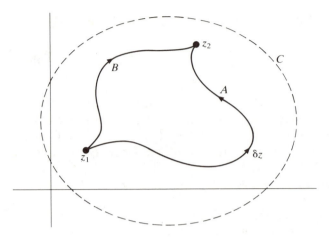

Figure P6-2 Contours in the z plane. If $f(z)$ is regular inside C, then $\int_{z_1}^{z_2} f(z)\, dz$ will give the same value if integrated along curve A as it will along contour B.

story. Take the example of $f(z) = 1/z$, and a path from $z = 1$ anticlockwise around the 'unit circle' $|z| = 1$ back to $z = 1$. We can spot that the general integral is $\log(z)$, and so the solution appears to be $\log(1) - \log(1)$. Unfortunately $\log(z)$ is not a single-valued function.

By writing z as $r \exp(j\theta)$, we see at once that (Fig. P6-3)

$$\log(z) = \log(r) + j\theta$$

As we take z around the unit circle, θ increases until by the time we have arrived back at $z = 1$, $\theta = 2\pi$. We could take another trip around the circle to increase θ to 4π, and so on. With anticlockwise or clockwise circuits, θ can take any value $2n\pi$, where n is an integer positive or negative.

Clearly θ will clock up a change of 2π if z takes any trip around the origin, the singularity, but for a journey around a small local circuit θ will return to the same value.

In the case of $f(z) = 1/z$, the integral in question will have the value $2\pi j$. For $f(z) = 7/z$, the integral would be $14\pi j$. For $f(z) = \cos(z)/z$, the integral would have the value $2\pi j$—since $\cos(0) = 1$.

In general, the integral around a simple pole at $z = a$ will be $2\pi j$ times the 'residue' at the pole, that is the value of $(z - a)\, f(z)$ as z tends to a. The residue can be thought of as the value of the function when the offending factor of the denominator is left out.

We can turn this around the other way and say that if $f(z)$ is 'regular' inside the unit circle (or any other loop for that matter), then we can find out the value of $f(a)$ (a is inside the loop) by integrating $f(z)/(z - a)$ around the loop to obtain $2\pi j\, f(a)$. It seems to be doing things the hard way, but it has its uses.

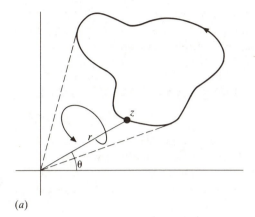

(a)

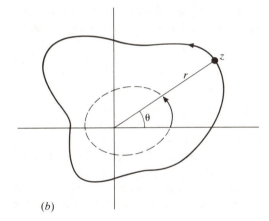

(b)

Figure P6-3 Evaluating $\log(z)$ around a contour. (a) θ returns to its original value. (b) θ changes by 2π.

Before we leave this topic, let us establish some facts for a later chapter. Integrating $1/z$ around the unit circle gives the value $2\pi j$. Integrating $1/z^2$ gives the value zero. Its general integral is $-1/z$, and this is no longer multivalued. In fact, the integral of z^n around the circle for any integer n will be zero, except for $n = -1$.

If $f(z)$ can be represented by a power series in z,

$$f(z) = \sum a_n z^{-n}$$

then we can pick out the coefficient a_n by multiplying $f(z)$ by z^{n-1} and integrating around the unit circle. This gives us a *transform* between sequences of values a_n and a function of z. We can construct the function of z simply by multiplying the coefficients by the appropriate powers of z and summing (all right, there is the question of convergence), and we can unravel the coefficients again by the integral method.

This transform between a sequence of values and a function of z will be of great value when we consider sampled systems. There are two other transforms of more immediate interest, Fourier and Laplace.

P6-5 DIFFERENTIAL EQUATIONS AND THE LAPLACE TRANSFORM

In Secs P2-1 and 2-4 we have glanced at the problems of finding analytic solutions to differential equations. The D operator method was mentioned, accompanied by warnings that there were limits to the liberties that could be taken with it. Now we will look at an alternative approach which will give a rigorous solution.

Suppose we have a function of time, $f(t)$, and that we are really only interested in its values for $t > 0$. For reasons we will see later, we can multiply $f(t)$ by the exponential function $\exp(-st)$ and integrate from $t = 0$ to infinity. We can write the result as

$$F(s) = \int_{t=0}^{\infty} f(t)\, e^{-st}\, dt$$

The result is written as $F(s)$, because since the integral has been evaluated over a defined range of t, t has vanished from the resulting expression. On the other hand, the result will depend on the value chosen for s.

In the same way that we have been thinking of $f(t)$ not as just one value but as a graph of $f(t)$ plotted against time, so we can consider $F(s)$ as a function defined over the entire range of s. We have transformed the function of time, $f(t)$, into a function of the variable s. This is the unilateral Laplace transform.

Consider an example. The Laplace transform of the function $5e^{-at}$ is given by

$$\int_{t=0}^{\infty} 5e^{-at} \times 5e^{-st}\, dt = \int_{t=0}^{\infty} 5e^{-(s+a)t}\, dt$$

$$= -\frac{5}{s+a}\, [e^{-(s+a)t}]_{t=0}^{\infty}$$

$$= \frac{5}{s+a}$$

We have a transform that turns functions of t into functions of s. Can we work the trick in reverse? Given a function of s, can we find just one function of time of which it is the transform? We may or may not arrive at a precise mathematical process for finding the 'inverse'—it is sufficient to spot a suitable function, provided that we can show that the answer is unique.

For 'well-behaved' functions, we can show that the transform of the sum of two functions is the sum of the two transforms:

$$\mathcal{L}\{f(t) + g(t)\} = \int_{t=0}^{\infty} [f(t) + g(t)] \, e^{-st} \, dt$$

$$= \int_{t=0}^{\infty} f(t) \, e^{-(s+a)t} \, dt + \int_{t=0}^{\infty} g(t) \, e^{-(s+a)t} \, dt$$

$$= F(s) + G(s)$$

Now suppose that two functions of time, $f(t)$ and $g(t)$, have the same Laplace transform. Then the Laplace transform of their difference must be zero:

$$\mathcal{L}\{f(t) - g(t)\} = \int_{t=0}^{\infty} [f(t) - g(t)] \, e^{-st} \, dt$$

$$= F(s) - G(s)$$

$$= 0$$

since we have assumed $F(s) = G(s)$. What values can $[f(t) - g(t)]$ take, if its Laplace integral is zero for every value of s? It can be shown that if we require $f(t) - g(t)$ to be differentiable, then it must be zero for all $t > 0$; in other words the inverse transform (if it exists) is unique.

Why should we be interested in leaving the safety of the time domain for these strange functions of s? Consider the transform of the derivative of $f(t)$:

$$\mathcal{L}\{f'(t)\} = \int_{t=0}^{\infty} f'(t) \, e^{-st} \, dt$$

Integrating by parts, we see

$$\mathcal{L}\{f'(t)\} = [f(t) \, e^{-st}]_{t=0}^{\infty} - \int_{t=0}^{\infty} f(t)(-s) \, e^{-st} \, dt$$

$$= -f(0) + s \int_{t=0}^{\infty} f(t) \, e^{-st} \, dt$$

$$= sF(s) - f(0)$$

We can use this result to show that

$$\mathcal{L}\{f''(t)\} = s\mathcal{L}\{f'(t)\} - f'(0)$$

$$= s[sF(s) - f(0)] - f'(0)$$

$$= s^2 F(s) - sf(0) - f'(0)$$

and in general

$$f^{(n)}(t) = s^n F(s) - s^{n-1}f(0) - s^{n-2}f'(0) - \cdots - f^{(n-1)}(0)$$

So what is the relevance of all this to the solution of differential equations? Suppose we are faced with

$$\ddot{x} + x = 5e^{-at}$$

Now if $\mathcal{L}\{x(t)\}$ is written as $X(s)$, we have

$$\mathcal{L}\{\ddot{x}(t)\} = s^2 X(s) - sx(0) - \dot{x}(0)$$

so

$$\mathcal{L}\{\ddot{x} + x\} = (s^2 + 1)X - sx(0) - \dot{x}(0)$$

We have already worked out that

$$\mathcal{L}\{5e^{-at}\} = \frac{5}{s + a}$$

so

$$(s^2 + 1)X - sx(0) - \dot{x}(0) = \frac{5}{s + a}$$

or

$$X(s) = \frac{1}{s^2 + 1}\left[\frac{5}{s + a} + sx(0) + \dot{x}(0)\right] \tag{P6-6}$$

Without too much trouble we have obtained the Laplace transform of the solution, complete with initial conditions. But how do we unravel the time function? Must we perform some infinite contour integration or other? Not a bit!

The art of the Laplace transform is to divide the solution into recognizable fragments. They are recognizable because we can match them against a table of transforms representing solutions to 'classic' differential equations. Some of the transforms might have been obtained by infinite integration, as we showed with e^{-at}, but others follow more easily by looking at differential equations.

The general solution to

$$\ddot{x} + x = 0$$

is

$$x = A\cos(t) + B\sin(t)$$

Now

$$\mathcal{L}\{\ddot{x} + x\} = 0$$

so

$$X(s) = \frac{1}{s^2 + 1}[sx(0) + \dot{x}(0)]$$

For the function $x = \cos(t)$, $x(0) = 1$ and $\dot{x}(0) = 0$, so

$$\mathscr{L}\{\cos(t)\} = \frac{s}{s^2 + 1}$$

If $x = \sin(t)$, $x(0) = 0$ and $x(0) = 1$, so

$$\mathscr{L}\{\sin(t)\} = \frac{1}{s^2 + 1}$$

With these functions in our table, we can settle two of the terms of Eq. (P6-6). We are left, however, with the term

$$\frac{1}{s^2 + 1} \frac{5}{s + a}$$

Using partial fractions, we can crack it apart into

$$\frac{A + Bs}{s^2 + 1} + \frac{C}{s + a}$$

Before we know it, we find ourselves having to solve simultaneous equations for A, B and C; these are equivalent to the equations we would have to solve for the initial conditions in the *classical* method.

Exercise P6-5-1   *Find the time solution of Eq. (P6-6) by solving for A, B and C as above and substituting back from the known transforms. Then solve the original differential equation the 'classic' way and compare the algebra involved.*

The Laplace transform really is not a magical method of solving differential equations. It is a systematic method of reducing the equations, complete with initial conditions, to a standard form, allowing the solution to be pieced together from a table of potted functions. Do not expect it to perform miracles, but do not underestimate its value.

P6-6 THE FOURIER TRANSFORM

After battling with Laplace, the Fourier transform may seem rather tame. The Laplace transformation involved the multiplication of a function of time by $\exp(-st)$ and its integration over all positive time. The Fourier transform requires the time function to be multiplied by $\exp(-j\omega t)$ and then integrated over all time past and future.

It can be regarded as the analysis of the time function into all its frequency components, which are then presented as a frequency spectrum. This spectrum contains phase information, so that by adding all the sinusoidal contributions the original function of time can be reconstructed.

Start by considering the Fourier series. This can represent a repetitive function as the sum of sine and cosine waves. If we have a waveform which repeats after time $2T$, it can be broken down into the sum of sinusoids of period $2T$, together with their harmonics.

Let us set out by constructing a repetitive function of time in this way. Rather than fight it out with sines and cosines, we can allow the function to be complex, and we can take the sum of complex exponentials $\exp(n\pi t/T)$:

$$f(t) = \sum_{n=-\infty}^{\infty} c_n \, e^{n\pi jt/T} \tag{P6-7}$$

Can we break $f(t)$ back down into its components? Can we evaluate the coefficients c_n from the time function itself?

The first thing to notice is that because of its cyclic nature, the integral of $\exp(n\pi jt/T)$ from $-T$ to $+T$ will be zero, for any integer n except zero. The integral will give

$$\frac{T}{n\pi j} \left[e^{n\pi jt/T} \right]_{-T}^{+T} = \frac{T}{n\pi j} \left(e^{n\pi j} - e^{-n\pi j} \right)$$

$$= 0$$

If $n = 0$, the exponential degenerates into a constant value of unity and the value of the integral will be just $2T$.

Now if we multiply $f(t)$ by $e^{-r\pi jt/T}$ we will have the sum of terms

$$f(t)\, e^{-r\pi jt/T} = \sum_{n=-\infty}^{\infty} c_n \, e^{(n-r)\pi jt/T}$$

If we integrate over the range $-T$ to $+T$, the contribution of every term on the right will vanish, except one. This will be the term where $n = r$, and its contribution will be $c_r . 2T$.

Now we have

$$c_r = \frac{1}{2T} \int_{-T}^{+T} f(t)\, e^{-r\pi jt/T} \, dt \tag{P6-8}$$

When we set this alongside (P6-7), we have a route from the coefficients to the time function and a return trip back to the coefficients.

We considered the Fourier series as a representation of a function which repeated over a period $2T$, i.e. where its behaviour over $-T$ to $+T$ was repeated again and again outside those limits. If we have a function that is not really repetitive, we can still match its behaviour in the range $-T$ to $+T$ with a Fourier series (Fig. P6-4).

The lowest frequency in the series will be π/T. The replica function will match over the centre range, but will diverge outside it. If we want to extend the range of the match, all we have to do is to increase the value of T. Suppose that we double it; then we will halve the lowest frequency present, and the interval between the frequency contributions will also be halved. In effect, the number of contributing frequencies in any range will be doubled.

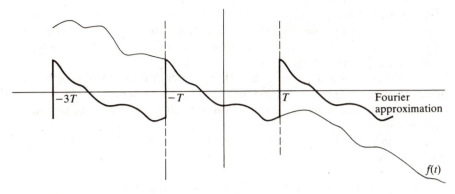

Figure P6-4 A Fourier series which matches the function over the interval $-T$ to T.

We can double T again and again, and increase the range of the match indefinitely. As we do so, the frequencies will bunch closer and closer until the summation of (P6-7) would be better written as an integral. But an integral with respect to what?

We are interested in frequency, which we write in angular terms as ω. The frequency represented by the (n)th term of (P6-7) is $n\pi/T$. Let us equate them. If we are going to integrate with respect to ω, then we need to define what we mean by $\delta\omega$. The obvious candidate is π/T. Now the summation of (P6-7) can be rewritten as

$$f(t) = \frac{1}{\pi} \sum_{\omega=n\pi/T} Tc(\omega)\, e^{j\omega t}\, \delta\omega \qquad (P6\text{-}9)$$

where we have written $c(\omega)$ for c_n and where (P6-8) has become

$$Tc(\omega) = \tfrac{1}{2} \int_{-T}^{+T} f(t)\, e^{-j\omega t}\, dt$$

Now $Tc(\omega)$ is an ungainly expression, especially as we intend to let T tend to infinity, so we will replace $2Tc(\omega)$ by $F(j\omega)$. We have in the limit, as T becomes infinite,

$$F(j\omega) = \int_{-\infty}^{\infty} f(t)\, e^{-j\omega t}\, dt \qquad (P6\text{-}10)$$

and

$$f(t) = \frac{1}{2\pi} \int_{-\infty}^{\infty} F(j\omega)\, e^{j\omega t}\, d\omega \qquad (P6\text{-}11)$$

We have a clearly defined way of transforming to the frequency domain and back to the time domain.

You may have noticed an amazing similarity between the integral of

(P6-10) and the integral defining the Laplace transform. Substitute s for $j\omega$ and they are closer still. They can be thought of as two variations of the same integral, where in the one case s takes only real values, while in the other its values are pure imaginary.

It will be of little surprise to find that the two transforms of a given time function appear algebraically very similar. Substitute $j\omega$ for s in the Laplace transform and you usually have the Fourier transform. They still have their separate uses.

This introduction sets the scene for the transforms. There is much more to learn about them, but that comes later.

Exercise P6-6-1 *Find the Fourier transform of $f(t)$, where*

$$f(t) = \begin{cases} \exp(-at) & \text{if } t > = 0 \\ 0 & \text{if } t < 0 \end{cases}$$

FREQUENCY DOMAIN METHODS

6-1 INTRODUCTION: THE PLOTS THICKEN

In the last chapter we saw the engineer testing a system to find how much feedback could be applied around it before instability set in. We saw the effectiveness of simple amplitude or power measurement to enable a frequency response to be plotted on log-log paper, and the development of rules of thumb into analytic methods backed by theory. We saw too that the output amplitude told only half the story; it was important also to measure the phase.

In the early days, phase measurement was not easy. The input and output waveforms could be compared on an oscilloscope, but the estimate of phase was somewhat rough and ready. If an $x-y$ oscilloscope was used, output could be shown as y against the input's x, enabling phase shifts of zero and multiples of 90° to be more accurately spotted (Fig. 6-1). Still the task of taking a frequency response was a tedious business.

Then came the introduction of the phase-sensitive voltmeter, often given the grand title of *transfer function analyser*. This contained the sine wave source to excite the system, and bore two large meters marked 'Reference' and 'Quadrature'. By using a *synchronous demodulator*, the box analysed the return signal into its components in-phase and 90° out-of-phase with the input—the real and imaginary parts of the complex amplitude, properly signed positive or negative.

It suddenly became easy to measure accurate values of complex gain, and the Nyquist diagram was straightforward to plot. With the choice of Nyquist, Bode, Nichols and Whiteley, control engineers could argue the benefits of

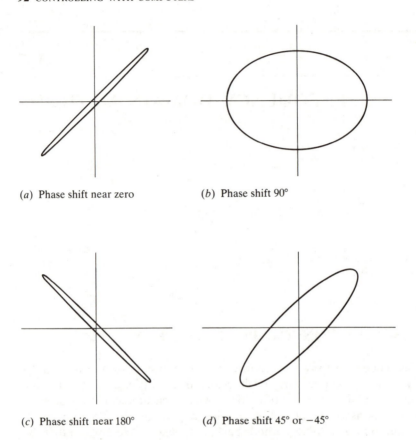

(*a*) Phase shift near zero

(*b*) Phase shift 90°

(*c*) Phase shift near 180°

(*d*) Phase shift 45° or −45°

Figure 6-1 Oscilloscope *x*–*y* traces for different phase shifts.

their particular favourite. Soon time domain and PRBS test methods were adding to the confusion—but they have no place in this chapter.

6-2 THE NYQUIST PLOT

Before looking at the variety of plots available, let us remind ourselves of the object of the exercise. We have a system which we believe will benefit from the application of feedback. Before 'closing the loop', we cautiously measure its open loop frequency response (or transfer function) to ensure that the closed loop will be stable. As a bonus, we would like to be able to predict the closed loop frequency response.

If the open loop transfer function is $G(s)$, the closed loop function will be

$$\frac{G(s)}{1 + G(s)} \tag{6-1}$$

since if the input is $U(s)$ and the output $Y(s)$, then the input to the inner system is $U(s) - Y(s)$. Now

$$Y(s) = G(s)[U(s) - Y(s)]$$

that is

$$[1 + G(s)]Y(s) = G(s)U(s)$$

so

$$Y(s) = \frac{G(s)}{1 + G(s)} U(s)$$

We saw that stability was a question of the location of the poles of a system, with disaster if any pole strayed to the right half of the complex frequency plane. Where will we find the poles of the closed loop system? Clearly they will lie at the values of s that give $G(s)$ the value -1. The complex gain $(-1 + j \cdot 0)$ is going to become the focus of our attention.

If we plot the readings from the phase-sensitive voltmeter, with the imaginary part against the real with no reference to frequency, we have a Nyquist plot. It is the path traced out in the complex gain plane as the variable s takes the values $j\omega$ and as ω increases from zero to infinity. It is the image in the complex gain plane of the positive part of the imaginary s axis.

Suppose that $G(s) = 1/(1 + s)$ (Fig. 6-2); then

$$G(j\omega) = \frac{1}{1 + j\omega}$$

$$= \frac{1 - j\omega}{1 + \omega^2}$$

$$= \frac{1}{1 + \omega^2} + j\frac{-\omega}{1 + \omega^2}$$

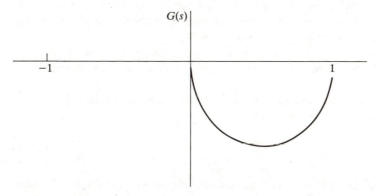

Figure 6-2 Nyquist plot of $1/(s + 1)$.

If G is plotted in the complex w plane as $u + jv$, then it is not hard to show that

$$u^2 + v^2 - u = 0$$

This represents a circle (for 'genuine' frequencies we can only plot a semicircle; the upper half of the circle is given by considering $s = -j\omega$) with a diameter formed by the line joining the origin to $(1 + j.0)$. What does this tell us about stability?

Clearly the gain drops to zero by the time the phase shift has reached 90° and there is no possible approach to the critical gain value of -1. Let us consider something more ambitious.

The system with the transfer function

$$G(s) = \frac{1}{s(s + 1)(s + 1)} \tag{6-2}$$

has a phase shift that is never less than 90° and approaches 270° at high frequencies, so it could have a genuine stability problem. We can substitute $s = j\omega$ and manipulate the expression to separate real and imaginary parts:

$$G(j\omega) = \frac{1}{j\omega} \frac{1}{1 - \omega^2 + 2j\omega}$$

$$= \frac{1}{\omega(1 + \omega^2)^2} [-j(1 - \omega^2) - 2\omega]$$

The imaginary part becomes zero at the value $\omega = 1$, leaving a real part with value $-\frac{1}{2}$. Once again, algebra tells us that there is no problem of instability. Suppose that we do not know the system in algebraic terms, but must base our judgement on the results of a frequency response. The Nyquist diagram is shown in Fig. 6-3. Just how much can we deduce from it?

From the intercept of -0.5 on the negative real axis, we can see that there is a gain margin of a factor of two. We can also check the phase margin by looking at the intersection of the trajectory with the unit circle, where the amplitude of the open loop response is unity.

We might wish to know the extent of any resonance in the closed loop system, i.e. the maximum gain we may expect. We can use the technique of M circles, as follows.

The closed loop gain $C(j\omega)$ is related to the open loop gain $G(j\omega)$ by the relationship

$$C = \frac{G}{1 + G} \tag{6-3}$$

Now C is an analytic function of the complex variable G, and the relationship supports all the honest-to-goodness properties of a mapping. We can show

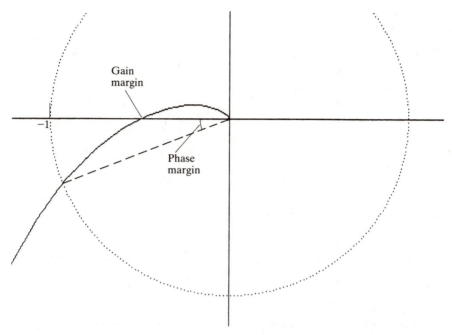

Figure 6-3 Nyquist plot of $1/s(s + 1)^2$.

an interest in the circles around the origin which represent various magnitudes of the closed loop gain, C. We can investigate the G plane to find out which curves map into the constant-magnitude C circles, and discover the answer to be another family of circles, as shown in Fig. 6-4.

Exercise 6-2-1 *By letting $G = x + jy$ and equating $|C|^2$ to m^2, use Eq. (6-3) to derive the equation of the locus of G.*

Again we see that we have a safely stable system, although the gain peaks at a value of 2.88.

We might be tempted to try a little more gain in the loop. We know that doubling the gain would put the curve right through the critical -1 point, so some smaller value must be used. Suppose we increase the gain by 50 per cent, giving an open loop gain of

$$G(s) = \frac{1.5}{s(s + 1)(s + 1)}$$

The Nyquist plot now sails rather closer to the critical -1 point, crossing the axis at -0.75, and the M circles show there is a resonance with a closed loop gain of 7.4 (Fig. 6-5). Can we deduce anything about the time response to a disturbance? Surprisingly, we can.

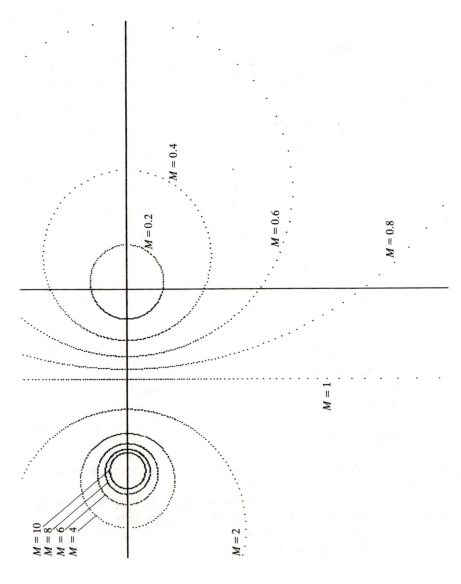

Figure 6-4 M circles.

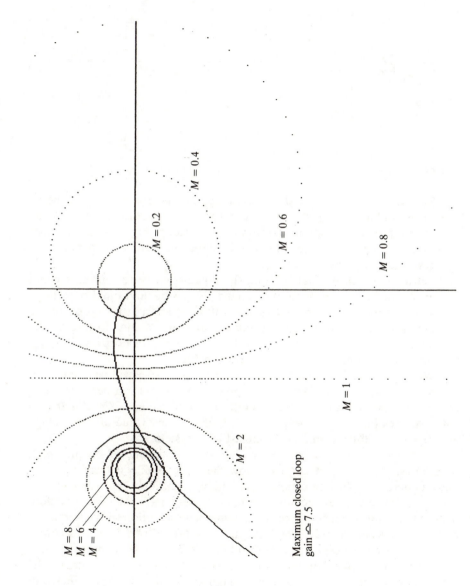

Figure 6-5 Nyquist plot of $1.5/s(s + 1)^2$ with M circles added.

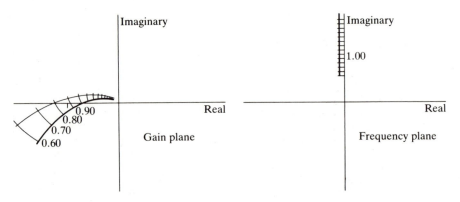

Figure 6-6 Mapping of a 'ladder' in the frequency domain onto the Nyquist plot of $1.5/s(s + 1)^2$.

Remember that the plot shows the mapping into the G plane of the $j\omega$ axis of the s plane. It is just one curve in the mesh which would have to be drawn to represent the 'mapped graph paper' effect of Fig. P6-1. Remember also that the squares of the coordinate grid of the s plane must map into 'curly squares' in the G plane.

Let us now tick off marks along the Nyquist curve to represent equal increments in frequency, say of 0.1 radians per second, and build onto these segments a mosaic of near-squares. We will have an approximation to the mapping not only of the imaginary axis but also of a 'ladder' formed by the imaginary axis, by the vertical line $-0.1 + j\omega$, and with horizontal 'rungs' joining them at intervals of 0.1j, as shown in Fig. 6-6.

We saw in Sec. 5-4 that the response to a disturbance will be made up of terms of the form $\exp(pt)$, where p is a pole of the overall transfer function—and in this case we are interested in the closed loop response. The closed loop gain becomes infinite only when $G = -1$, and so any value of s that maps into $G = -1$ will be a pole of the closed loop system.

Looking closely at the 'curved ladder' of our embroidered Nyquist plot, we see that the -1 point lies just below the 'rung' of $\omega = 0.9$ and just past midway across it. We can estimate reasonably accurately that the image of the -1 point in the s plane has coordinates $-0.055 + j \cdot 0.89$. We therefore deduce that a disturbance will result in a damped oscillation with a frequency of 0.89 radians per second and a decay time constant of $1/0.055 = 18$ seconds.

(If you are worried that poles should come in complex conjugate pairs, note that the partner of the pole we have found here will lie in the corresponding image of the negative imaginary s axis, the mirror image of this Nyquist plot, which is usually omitted for simplicity.)

Exercise 6-2-2 By drawing 'curly squares' on the plot of $G(s) = 1/[s(s + 1)^2]$ (Fig. 6-3) estimate the resonant frequency and damping factor for unity feedback. (Algebra gives $s = -0.122 + j \cdot 0.745$.)

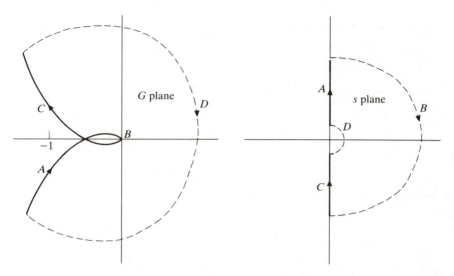

Figure 6-7 The Nyquist plot of $1/s(s + 1)^2$ completed to show the mapping of the positive frequency plane D.

With mappings in mind, we can be a little more specific about the conditions for stability. We can regard the plot not just as the mapping of the positive imaginary s axis but of a journey outwards along the axis. As s moves outwards, so in Exercise 6-2-2 does G move from values in the lower left quadrant in towards the G plane origin. As it does so, G leaves the -1 point on its left-hand side, implying from the theory of complex functions that s leaves the corresponding pole on the left-hand side of the imaginary axis—the safe side.

We can extend this concept by considering a journey in the s plane upwards along the imaginary axis to a very large value, then in a clockwise semicircle enclosing much of the 'dangerous' half-plane, then back up the negative imaginary axis to the origin. In making such a journey, we must not encircle any poles if the system is to be stable. This results in the requirement that the 'completed' G curve must not encircle the -1 point.

When G becomes infinite at $s = 0$, as in our present example, we can bend the journey in the s plane to make a small anticlockwise semicircular detour around $s = 0$, as shown in Fig. 6-7.

6-3 NYQUIST SUMMARY

We have seen a method of testing an unknown system and plotting the in-phase and quadrature parts of the open loop gain to give an insight into closed loop behaviour. We have not only a test for stability, by checking to

```
5 SCREEN 3:CLS
10 NX=50:TDEC=10^(3/NX):REM Frequency increments log scale
15 REM                         over three decades of freq.
20 GRSC=200:GISC=165:GR0=300:GI0=100:PI=3.14159
25 REM Scales for real, imaginary, Datum for origin
30 INPUT"how many poles ";NP:DIM P(NP,1)
40 FOR I=1 TO NP:INPUT"real, imag ";P(I,0),P(I,1):NEXT
50 INPUT"how many zeroes ";NZ:DIM Z(NZ,1)
60 FOR I=1 TO NZ:INPUT"real, imag ";Z(I,0),Z(I,1):NEXT
70 INPUT"Gain factor    ";G0  :CLS
80 LINE(0,GI0)-(600,GI0):REM Real axis
85 LINE(GR0,0)-(GR0,500):REM Imaginary axis
90 FOR N=0 TO 2*PI STEP .02:PSET(GR0+GRSC*COS(N),GI0+GISC*SIN(N)):NEXT
95 REM: Dot in the unit circle
100 W=.1:S=0:FOR X=0 TO NX
110 GOSUB 200:REM Calculate complex gain GR + j GI
140 IF X=0 THEN PSET(GR0+GRSC*GR,GI0-GISC*GI)
150 LINE -(GR0+GRSC*GR,GI0-GISC*GI):W=W*TDEC
160 NEXT:END
200 NR=G0:NI=0:DR=1:DI=0
210 FOR I=1 TO NP:VR=S-P(I,0):VI=W-P(I,1):DR1=DR*VR-DI*VI
220 DI=DR*VI+DI*VR:DR=DR1:NEXT:MD=DR*DR+DI*DI
230 IF NZ=0 THEN 260
240 FOR I=1 TO NZ:VR=S-Z(I,0):VI=W-Z(I,1):NR1=NR*VR-NI*VI
250 NI=NR*VI+NI*VR:NR=NR1:NEXT
260 GR=(NR*DR+NI*DI)/MD:GI=(NI*DR-NR*DI)/MD
270 RETURN
```

(*a*) Nyquist only

```
5 SCREEN 3:CLS
10 NX=50:TDEC=10^(3/NX):REM Frequency increments log scale
15 REM                         over three decades of freq.
20 GRSC=200:GISC=165:GR0=300:GI0=100:PI=3.14159
25 REM Scales for real, imaginary, Datum for origin
30 INPUT"how many poles ";NP:DIM P(NP,1)
40 FOR I=1 TO NP:INPUT"real, imag ";P(I,0),P(I,1):NEXT
50 INPUT"how many zeroes ";NZ:DIM Z(NZ,1)
60 FOR I=1 TO NZ:INPUT"real, imag ";Z(I,0),Z(I,1):NEXT
70 INPUT"Gain factor    ";G0  :CLS
80 LINE(0,GI0)-(600,GI0):REM Real axis
85 LINE(GR0,0)-(GR0,500):REM Imaginary axis
90 FOR M=.2 TO 1 STEP .2:GOSUB 500:NEXT :REM M-circles
95 FOR M=2 TO 10 STEP 2:GOSUB 500:NEXT
100 W=.1:S=0:FOR X=0 TO NX
110 GOSUB 200:REM Calculate complex gain GR + j GI
140 IF X=0 THEN PSET(GR0+GRSC*GR,GI0-GISC*GI)
150 LINE -(GR0+GRSC*GR,GI0-GISC*GI):W=W*TDEC
160 NEXT:END
200 NR=G0:NI=0:DR=1:DI=0
210 FOR I=1 TO NP:VR=S-P(I,0):VI=W-P(I,1):DR1=DR*VR-DI*VI
220 DI=DR*VI+DI*VR:DR=DR1:NEXT:MD=DR*DR+DI*DI
230 IF NZ=0 THEN 260
240 FOR I=1 TO NZ:VR=S-Z(I,0):VI=W-Z(I,1):NR1=NR*VR-NI*VI
250 NI=NR*VI+NI*VR:NR=NR1:NEXT
260 GR=(NR*DR+NI*DI)/MD:GI=(NI*DR-NR*DI)/MD
270 RETURN
500 FOR N=.05 TO 2*PI-.05 STEP .04: X=COS(N): Y=SIN(N)
510 NR=-M*X:NI=-M*Y:DR=M*X-1:DI=M*Y:MD=DR*DR+DI*DI
520 GR=(NR*DR+NI*DI)/MD:GI=(NI*DR-NR*DI)/MD
530 PSET(GR0+GRSC*GR,GI0-GISC*GI):NEXT:RETURN
```

(*b*) With *M* circles added

Figure 6-8 Listing of BASIC programs for Nyquist plots. (a) Nyquist only. (b) With M-circles added. (*Note*: if SCREEN 3 is not available on your machine, edit line 5 to use SCREEN 2 or SCREEN 1. You will then need to adjust the scale factors in line 20.)

see if the -1 point is passed on the wrong side, but an accurate way of measuring the peak gains of resonances. What is more, we can in many cases extend the plot by 'curly squares' to obtain an estimate of the natural frequency and damping factor of a dangerous pole.

This is all performed in practice without a shred of algebra, simply by plotting the readings of an 'R and Q' meter on linear graph paper, estimating closed loop gains with the aid of preprinted M circles.

However straightforward it might be to plot experimental readings on a Nyquist diagram, it is still handy to let the computer draw a theoretical plot. The program listed in Fig. 5-7 can be somewhat simplified, resulting in the listing of Fig. 6-8. Verify the Nyquist diagrams in this chapter.

6-4 THE NICHOLS CHART

The R & Q meter lent itself naturally to the plotting of Nyquist diagrams, but suppose that the gain data was obtained in the more 'traditional' form of gain and phase, as used in the Bode diagram. Is it meaningful to plot the logarithmic gain directly against the phase angle, and what could be the advantages?

We have seen that an analytic function has some useful mathematical properties. It can also be shown that an analytic function of an analytic function is itself analytic. The logarithm function is a good honest analytic function, where

$$\log[G(s)] = \log(|G|) + j \arg(G)$$

(Remember that the function *arg* gives the phase angle in radians, with the value $\tan^{-1}[\mathrm{Imag}(G)/\mathrm{Real}(G)] + n\pi$, where n puts the result into the correct quadrant of the complex plane.)

Instead of plotting the imaginary part of G against the real, as for Nyquist, we can plot the logarithmic gain in decibels against the phase shift of the system. All the rules about encircling the critical point where $G = -1$ will still hold, and we should be able to find the equivalent of M circles.

The point $G = -1$ will of course now be defined by phase shift $= 180°$ together with gain $= 0$ dB. The M circles are circles no longer, but since the curves can be preprinted onto the chart paper that is no great loss. The final effect is shown in Fig. 6-9.

So where does any advantage lie?

Suppose that we wish to consider a variety of gains applied at the open loop transfer function input, as defined by k in Fig. 6-10. To estimate the resulting closed loop response, we might have to rescale a Nyquist diagram for each value considered. Changing the gain in a Nichols plot, however, is simply a matter of moving the plot vertically with respect to the chart paper.

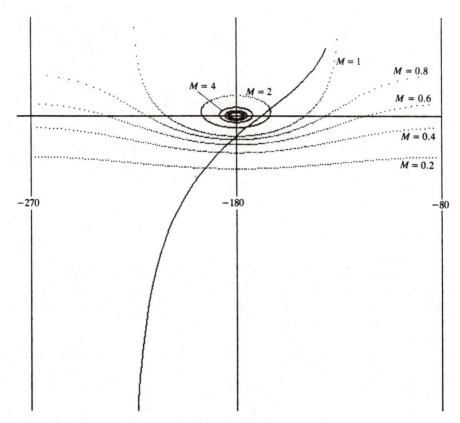

Figure 6-9 Nichols chart for $1/s(s + 1)^2$. The dotted curves are the equivalent of M circles.

By inspecting Fig. 6-9 it is not hard to estimate the value of k, for instance, that will give a peak closed loop gain of 3.

When more sophisticated compensation is considered, such as phase advance, the Nichols chart relates closely to the Bode diagram and the two can be used together to good effect.

To plot the Nichols diagram requires only a slight modification of the program listed in Fig. 5-5. Now grid lines are drawn for phase shifts of 0°, 180° and 360° and for unity gain. The same subroutine is used to calculate the

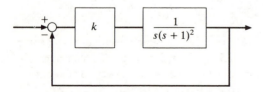

Figure 6-10 Unity feedback around a variable gain.

```
5 SCREEN 3:CLS
10 NX=50:TDEC=10^(3/NX):REM Frequency increments log scale
15 REM                     over three decades of freq.
20 GSC=30:PHSC=80:PHØ=600:LGØ=100:PI=3.14159
25 REM Scales for X, gain, phase, Datum for phase, gain
30 INPUT"how many poles ";NP:DIM P(NP,1)
40 FOR I=1 TO NP:INPUT"real, imag ";P(I,Ø),P(I,1):NEXT
50 INPUT"how many zeroes ";NZ:DIM Z(NZ,1)
60 FOR I=1 TO NZ:INPUT"real, imag ";Z(I,Ø),Z(I,1):NEXT
70 INPUT"Gain factor      ";GØ :CLS
80 LINE(Ø,LGØ)-(600,LGØ):REM Unity gain line
85 LINE(PHØ,Ø)-(PHØ,500):REM Phase = Ø
90 LINE(PHØ-PHSC*PI,Ø)-(PHØ-PHSC*PI,500):REM Phase = -PI
95 LINE(PHØ-PHSC*2*PI,Ø)-(PHØ-PHSC*2*PI,500):REM Phase = -2*PI
100 W=.1:S=Ø:FOR X=Ø TO NX
110 GOSUB 200:REM Calculate complex gain GR + j GI
120 LG=LOG(GR*GR+GI*GI)/2:REM Log Gain (base e)
130 PH=ATN(GI/GR):IF GR<Ø THEN PH=PH-PI
140 IF X=Ø THEN PSET(PHØ+PHSC*PH,LGØ-GSC*LG)
150 LINE -(PHØ+PHSC*PH,LGØ-GSC*LG):W=W*TDEC
160 NEXT:END
200 NR=GØ:NI=Ø:DR=1:DI=Ø
210 FOR I=1 TO NP:VR=S-P(I,Ø):VI=W-P(I,1):DR1=DR*VR-DI*VI
220 DI=DR*VI+DI*VR:DR=DR1:NEXT:MD=DR*DR+DI*DI
230 IF NZ=Ø THEN 260
240 FOR I=1 TO NZ:VR=S-Z(I,Ø):VI=W-Z(I,1):NR1=NR*VR-NI*VI
250 NI=NR*VI+NI*VR:NR=NR1:NEXT
260 GR=(NR*DR+NI*DI)/MD:GI=(NI*DR-NR*DI)/MD
270 RETURN
```

Figure 6-11 Listing of a BASIC program for Nichols plots. (*Note:* If SCREEN 3 is not available on your machine, edit line 5 to use SCRREEN 2 or SCREEN 1. You will then need to adjust the scale factors in line 20.)

complex gain and the same code extracts the logarithm of the gain and its phase angle. Figure 6-11 shows the new listing. In line 150 these are plotted against each other, rather than against frequency.

Make the modifications, including the scale factors of line 20, and verify the shape of Fig. 6-9.

6-5 THE INVERSE NYQUIST DIAGRAM

There is an alternative to the Nyquist diagram which maintains simplicity while making it easy to relate open loop to closed loop gain. It is sometimes called the Whiteley diagram.

We have become accustomed to think in terms of gain—to apply one volt to the input of a circuit and measure the output. An equally valuable concept is inverse gain; what input voltage will give an output of just one volt? When we start closing loops, we see that inverse gain is much simpler to deal with. Consider a system with open loop gain G and unity feedback, as in Fig. 6-12. If the output is one unit, then the input to the G element is $1/G$. The

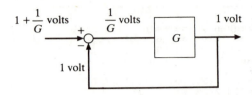

Figure 6-12 Calculations based on inverse gain.

feedback is again one unit, so the input to the closed loop system is $1 + 1/G$. If we use the symbol W for inverse gain, then

$$W_{\text{closed loop}} = W_{\text{open loop}} + 1$$

That is really all there is to it. The frequency response of the closed loop system is obtained from the open loop just by moving it one unit to the right. In the open loop plot, the -1 point is still the focus of attention—it moves to the origin when the loop is closed, and the origin represents one unit of output for zero input. The M circles are still circles, but now they are simply centred on the -1 point. Radius 2 implies a closed loop gain of $\frac{1}{2}$, and so on. Phase shift is equally simply given by measuring the angle of the vector joining the point in question to the -1 point—and then negating the answer.

Some of the more familiar transfer functions become almost ridiculously easy. The lag,

$$G(s) = \frac{1}{1+s}$$

which gave a semicircular Nyquist plot, now appears as

$$W(s) = 1 + s$$

When we give s a range of values $j\omega$, the Whiteley plot is simply a line rising vertically from the point $W = 1$. We close the loop with unity feedback, and see that the line has moved one unit to the right to rise from $W = 2$ (see Fig. 6-13).

With more complicated functions, we might become worried about the 'rest' of the plot, for negative frequencies and for the large complex

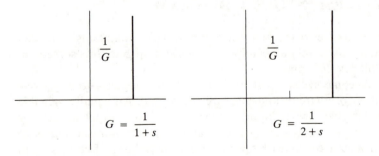

Figure 6-13 Whiteley plots for a simple system.

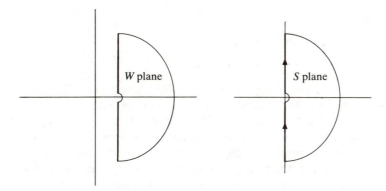

Figure 6-14 Completing a Whiteley plot for the frequency plane D.

frequencies that complete the loop in the s plane. With Nyquist, we usually have no need to bother, since the high frequency gain generally drops to zero and the plot muddles gently around the origin, well away from the -1 point. The inverse gain often becomes infinite, however, and the plot may soar around the boundaries of the plot.

In the $1/(1 + s)$ example, $W(s)$ approximates to s for large values of s, and so as s makes a clockwise semicircular detour around the positive real half-plane, so W will make a similar journey well away from the -1 point, as shown in Fig. 6-14.

It is now clear that the inverse Nyquist plot is at its best when we wish to consider a variable gain k in the feedback loop (see Fig. 6-15). Since the closed-loop gain is now the inverse of $A/G + k$ we can slide the Whiteley plot any distance to the right to examine any particular value of k. The stability tests will now apply to the point $W = -k$ in just the same manner as if we stay with $k = 1$. (With k in front of the G block as before, the stability and resonance tests are just the same, but we must allow for an extra factor of k in the closed loop gain given by the M circles.)

The damped motor found in examples of position control has a transfer function of the form

$$\frac{1}{s(s + a)}$$

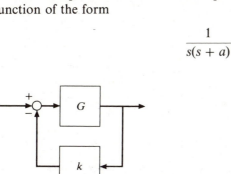

Figure 6-15 Variable feedback around a system.

so the inverse gain is

$$W(j\omega) = j\omega(j\omega + a)$$
$$= -\omega^2 + ja\omega$$

The plot is a simple parabola. The portion of the plot for negative frequencies completes the symmetry of the parabola, leaving us only to worry about very large complex frequencies. Now for large s, W approximates to s^2 and so as s makes its return journey from $+jR$ clockwise around a large semicircle via $+R$ to $-jR$, W will move from $-R^2$ clockwise through double the angle via $+R^2$ and round through negative imaginary values to $-R^2$ again. The entire plot is shown in Fig. 6-16. Clearly since it does not anywhere cut the real axis to the left of the -1 point, no amount of feedback will result in instability. The M circles show, however, that there can be a peak of resonance if k is made large.

To obtain an illustration that could go unstable, we have been using the third-order system

$$G(s) = \frac{1}{s(s + 1)^2}$$

What does this look like as an inverse Nyquist diagram?

$$W(j\omega) = j\omega(1 + j\omega)^2$$
$$= -2\omega^2 + j\omega(1 - \omega^2)$$

The plot starts from the origin, first taking values that have negative real part and positive imaginary, passes above the -1 point and cuts the negative real axis at $W = -2$. It then dives off south-west on a steepening curve. When we plot the values for negative frequency, it looks suspiciously as though the -1

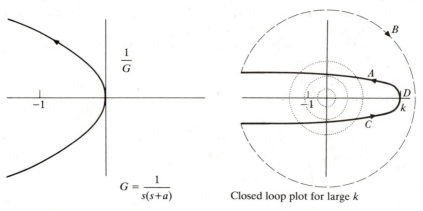

$$G = \frac{1}{s(s+a)}$$

Closed loop plot for large k

Figure 6-16 Whiteley plot of $1/s(s + a)$. Closed loop plot shows that heavy feedback can cause resonance, but not instability.

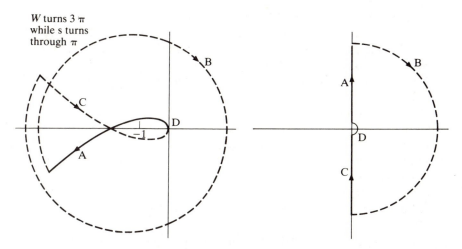

W turns 3 π
while s turns
through π

Figure 6-17 Whiteley diagram for $1/s(s + 1)^2$, completed, showing that despite first appearances the -1 point is not encircled.

point has been surrounded—disaster! We must, however, take great care to examine the return semicircular journey.

W now looks like the cube of *s* for large values, and so as *s* moves around its semicircular π radians, *W* will move in the same sense through 3π radians. When *s* is 'at the top' of the imaginary axis, *W* has a large negative imaginary part and a somewhat smaller negative real part. As *s* rotates clockwise, so *W* rotates clockwise through one-and-a-half revolutions to join the negative frequency plot high up and rather to the left of the imaginary axis, as shown in Fig. 6-17. If the plot were a loop of cotton and a pencil were stabbed into the -1 point, then the loop could be pulled clear without wrapping around the pencil. Despite first appearances the -1 point is not encircled, and the system is stable.

If we considered a feedback gain of 3, however, we would find ourselves looking at the point $W = -3$. This is well and truly encircled, twice in fact, and the system would be unstable.

6-6 SUMMARY OF EXPERIMENTAL METHODS

It must be repeated that the methods of these last two chapters relate to experimental testing of an unknown system. Although most of the examples have been based on simple transfer functions, this is merely for convenience of explanation and understanding. In practice the engineer might be presented with a jumbled table of readings from which to make an intelligent decision.

In the case of the Bode plot, an attempt is made to recognize breakpoints so that a guess can be made at the transfer function. Gain variations can be considered by simply raising or lowering the log–gain plot, and rule-of-thumb will get the engineer quite a long way. Even without recognizing the transfer function, some quite sophisticated compensators can be prescribed, in the form of phase advance or lag lead filters. Although stability is well defined in terms of gain margin or phase margin, however, prediction of the performance of the closed loop system is not easy.

The Nyquist plot takes the output from an R & Q meter and displays it quite simply as the imaginary part of gain against real. *M* circles enable the peak closed loop gain to be accurately predicted. Gain values for marginal stability are easily deduced, too, although deducing a closed loop response for anything but unity feedback involves some work. There are rigorous rules to deduce the stability of complicated compensators which cause the gain to exceed unity at a phase shift of 180°, yet pull the phase back again to avoid encircling the −1 point.

The Nichols plot is most simply applied to readings in the form of log gain and phase angle. It offers all the benefits of Nyquist, barring the distortion of the *M* curves, and allows easy consideration of variations in gain. When contemplating phase advance and other compensators, it makes a good partner to a Bode plot.

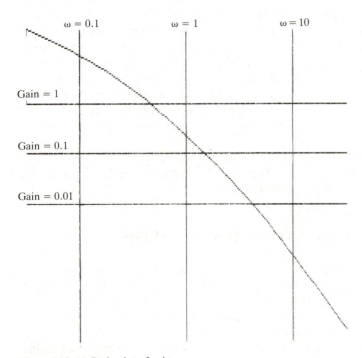

Figure 6-18 (a) Bode plot of gain.

The inverse Nyquist plot will usually require some calculation in order to turn the readings into plotted points; when a computer is involved, this is no handicap. It offers the same tests and criteria as Nyquist, but gives a much clearer consideration to variable feedback. Some head-scratching may be involved in deducing the 'closing' plot for large complex frequencies.

Exercise 6-6-1 *Figure 6-18a shows a Bode plot of amplitude alone. Make a guess at the phase plot. In Fig. 6-18b is shown a Nyquist plot. Can these be taken from the same system? Draw the corresponding Nichols plot. What is the maximum gain for stability? Estimate the transfer function.*

Exercise 6-6-2 *Draw an inverse Nyquist plot for the system*

$$G(s) = \frac{1}{s^2(1 + s)}$$

Can it be stabilized by proportional feedback alone?

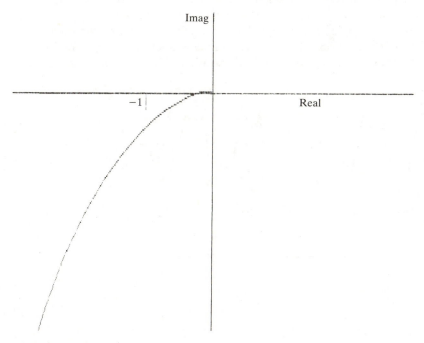

(b) Nyquist plot.

Whenever we have defined an example by its transfer function, there has been the temptation to bypass the graphics and manipulate the algebra to solve for resonant frequencies, limiting gains and so on. In the next chapter we will see how to take full advantage of such knowledge, and will look at the stability not just of one particular feedback gain but of an entire range so that we can make an intelligent choice.

6-7 NON-LINEARITY AND THE DESCRIBING FUNCTION

At the end of Chapter 3 we gave passing consideration to the stability of non-linear systems, looking for Liapunov functions that would aid us in diagnosisng the behaviour of trajectories in the state space. What corre-sponding techniques are open to us in the frequency domain?

By now we have become adept at looking for a loop gain of unity. We sought a signal which when passed through the open loop system would deliver up a signal for feedback exactly the same as the one we started with.

Let us state the obvious again: when we close the loop, the feedback signal arriving at the input is exactly the same as the input that produces it (Fig. 6-19).

To the signal at the input, the loop gain is exactly one, by any reckoning, whether it is a decaying exponential or a saturated square wave oscillation. Once we allow the system to become non-linear, the *eigenfunction* is no longer a simple (or complex!) exponential, but can take a variety of distorted forms.

To put a label onto the analysis of such a function, we must make some assumptions and apply some limitations. We will look mainly at such effects

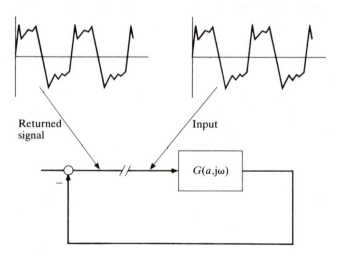

Figure 6-19 The returned signal is identical with the input.

as clipping, friction and backlash, and we will assume that the oscillation which we are guarding against is at least approximately sinusoidal.

With the assumption that the oscillation signal is one that repeats regularly, we open up the possibility of breaking the signal into a series of sinusoidal components by calculating its Fourier series. The fundamental sine wave component of the signal entering the system must be exactly equal to the fundamental component of the feedback, and so we can at least start to build up an equation.

We can consider the application of sine waves of varying amplitudes to the system, as well as of varying frequencies, and will extract a fundamental component from the feedback which is now multiplied by a gain function of both frequency and amplitude, the *describing function* of the system $G(a, j\omega)$. As ever, we are concerned with finding if G can take the value -1 for any combination of frequency and amplitude.

Of course, the method depends on an approximation. It ignores the effect of higher harmonics combining to recreate a signal at the fundamental frequency. However, this effect is likely to be small. We can use the method both to estimate the amplitude of oscillation in an unstable system that has constraints and to find situations where an otherwise stable system can be provoked into a limit cycle oscillation.

SEVEN

THE ROOT LOCUS

7-1 INTRODUCTION

In the last chapter, we saw that a reasonably accurate frequency response allowed us to plan the feedback gain to achieve a specified resonance peak. We could even make an educated guess at some of the poles of the closed loop system. If we have deduced or been given the transfer function in algebraic terms, however, we can use some more algebra to compute the transfer function of the closed loop system and to deduce the values of all its poles and zeros.

Faced with an intimidating list of eight-figure complex numbers, however, we might still not find the choice of feedback gain to be very easy. What we need is some way of visualizing in graphic terms the effect of varying the feedback; if we can avoid having to call upon the services of a computer, that will be an added bonus. What we need is the root locus plot.

7-2 ROOT LOCUS AND MAPPINGS

When examining an inverse Nyquist diagram, we saw that one or more of the closed loop poles could be assessed from the location of the -1 point within the 'curly squares'. In order to examine the effect of feedback gain k, we instead looked at the point $-k + j \cdot 0$. This also had a measurable position within one of the 'curly squares' which represent the mapping of the s plane 'graph paper' onto the plane of complex inverse gain. As k varies, we can trace out the path of the negative real inverse gain axis among the curly squares and make a choice of value for one of the closed loop poles, as shown in Fig. 7-1.

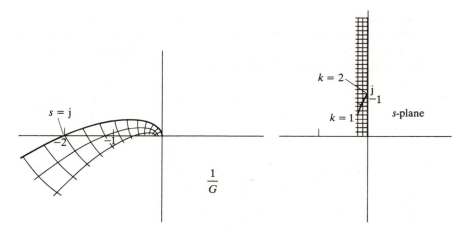

Figure 7-1 Whiteley diagram of $1/s(s + 1)^2$, showing points on the negative real gain axis mapped back into complex frequencies.

While we can estimate the most important poles, the pair in danger of becoming unstable, we have no clue about the third. The curly squares are awkward and inaccurate to draw, and could become very tangled for a more complicated system. There must be an easier way!

In struggling with the curly squares, we were trying to estimate the 'inverse mapping' of the -1 point, the point in the s plane which mapped into a gain of -1. Then we gazed earnestly at the inverse mapping of a $-k$ point. If we could only find a way of locating the inverse mapping in the s plane of the entire real negative gain axis, we would have a map of every possible value of closed loop pole. We could then start to make some choices about k.

The inverse mapping in the s plane of the negative real gain axis is called the root locus.

As we will see, it is not too hard to sketch the root locus by hand. With the aid of a computer, there are many ways of plotting or displaying the locus. One which is time consuming to run yet simple to program and entertaining to watch is listed as a program in Appendix A. The computer screen is used to represent a portion of the s plane, each character position having a corresponding complex value of s. The real and imaginary parts of each such s are substituted into an equation which results in the evaluation of the open loop gain, real and imaginary parts.

At this stage we are not interested in the actual values, just in whether the real and imaginary parts are positive or negative. If the real and imaginary parts of the gain are both positive we print a symbol '+' on the screen at that s location. If the real part is negative and the imaginary part is positive we print '−'. If both parts are negative we print 'E' while if the real is positive and the imaginary is negative we print 'X'. There is thus a good contrast of light and heavy characters between s points resulting in a positive imaginary part of the gain and those with negative. The points resulting in zero imaginary parts will be clearly marked out as the boundary between these regions.

This contrast boundary will show the inverse mapping of the entire real gain axis. On a printout of the screen, it only remains to ink in the boundaries bordered by '—' and we have the root locus (Fig. 7-2). This method of

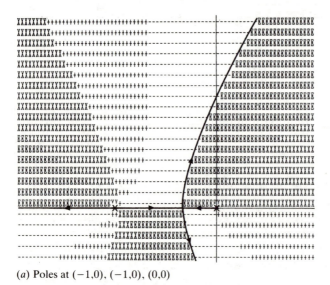

(*a*) Poles at $(-1,0)$, $(-1,0)$, $(0,0)$

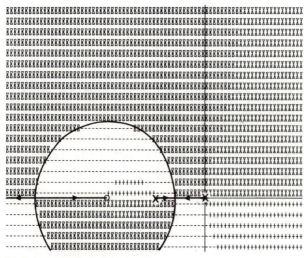

(*b*) Poles at $(0,0)$, $(-0.5,0)$
    Zero at $(-1,0)$

```
5 CLS
10 S0=-2 :S1=1:WM=2:REM svalues real, imag at edges
15 BLOB$="E-X+"
20 SS=(S1-S0)/79:SW=WM/20
30 INPUT"how many poles ";NP:DIM P(NP,1)
40 FOR I=1 TO NP:INPUT"real, imag ";P(I,0),P(I,1):NEXT
50 INPUT"how many zeroes ";NZ:DIM Z(NZ,1)
60 FOR I=1 TO NZ:INPUT"real, imag ";Z(I,0),Z(I,1):NEXT
100 W=-4.5*SW:FOR Y=23 TO 1 STEP -1
110 S=S0:FOR X=1 TO 79
200 NR=1:NI=0:DR=1:DI=0
210 FOR I=1 TO NP:VR=S-P(I,0):VI=W-P(I,1):DR1=DR*VR-DI*VI
220 DI=DR*VI+DI*VR:DR=DR1:NEXT
230 IF NZ=0 THEN 260
240 FOR I=1 TO NZ:VR=S-Z(I,0):VI=W-Z(I,1):NR1=NR*VR-NI*VI
250 NI=NR*VI+NI*VR:NR=NR1:NEXT
260 SR=NR*DR+NI*DI:SI=NI*DR-NR*DI
270 B=-2*(SR>0)-(SI>0)
300 LOCATE Y,X:PRINT MID$(BLOB$,B+1,1)
310 S=S+SS:NEXT
320 W=W+SW:NEXT
```

(c)

Figure 7-2 'Character plot' for root locus. (a) Plot for $1/s(s + 1)^2$. (b) Plot for $(s + 1)/s(s + 0.5)$. (c) Listing of program.

plotting is crude and of poor accuracy, but it gives a good insight into the essential nature of a plot which is sometimes treated with mystical reverence.

By now you may have realized that the paper cover of this book is embellished with a very similar design. Instead of a crude array of characters, however, the real and imaginary gain values are represented by printers' inks, red for positive real part and blue for positive imaginary.

7-3 POLES AND ZEROS

In Sec. 5-5 we suggested that a great many dynamic systems could be represented by a transfer function which is the ratio of two polynomials in s. Since any polynomial can be factorized into a product of terms of the form $(s - p_i)$ if the p's are allowed to be complex, we can write

$$G(s) = a \frac{(s - z_1)(s - z_2) \cdots (s - z_m)}{(s - p_1)(s - p_2) \cdots (s - p_n)} \tag{7-1}$$

If s takes the value of any of the poles, the value of G will be infinite. When s takes the value of one of the zeros, the gain is zero.

We can in fact use the method hinted at in Sec. 5-5 to work out the value of gain at any value of s. The magnitude of G is the product of the magnitudes of the numerator terms, divided by those of the denominator terms. It is not too difficult to extract a frequency response from an s plane plot of poles and

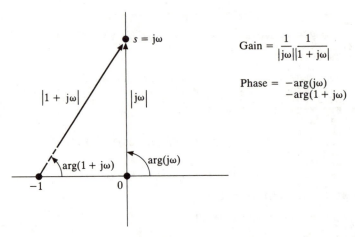

Figure 7-3 Gain and phase can be measured graphically from the vectors joining the selected point on the imaginary (frequency) axis to the poles and zeros.

zeros simply by measuring the lengths of the vectors joining poles and zeros to each chosen frequency point on the imaginary axis, and multiplying or dividing accordingly.

The argument (phase angle) of each frequency point is obtained by adding the arguments (angles anticlockwise from due east) of the vectors joining each of the zeros to the corresponding value of s and subtracting the sum of the arguments in the denominator. Some thirty years ago, there was a device on the market called a 'Spirule' which simplified the measurement and calculation of both amplitudes and angles.

Our concern now is not the open loop frequency response but the new locations of the poles when we close the loop with some feedback gain yet to be determined. Let us start by considering the example shown in Fig. 7-3:

$$G(s) = \frac{1}{s(s + 1)}$$

Suppose that this is the transfer function of a somewhat sluggish motor, which we wish to use in a position control loop as shown in Fig. 7-4.

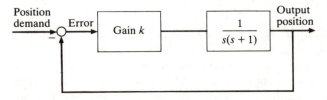

Figure 7-4 Position control with variable gain and unity feedback.

The closed loop gain is given by

$$\frac{kG}{1 + kG} = \frac{k/s(s + 1)}{1 + k/s(s + 1)}$$

$$= \frac{k}{s(s + 1) + k}$$

$$= \frac{k}{s^2 + s + k}$$

Now if we hammer this out, we will see that the roots are at

$$-\tfrac{1}{2} \pm \sqrt{\tfrac{1}{4} - k}$$

If k is very small, the roots are near -1 and 0, the open loop positions. As k increases toward $\tfrac{1}{4}$, the roots are still both real and lie symmetrically either side of $-\tfrac{1}{2}$. As k reaches $\tfrac{1}{4}$, the roots come together and form a double pole at $-\tfrac{1}{2}$. Then they become complex. As k increases, the roots move apart up and down the vertical line through $-\tfrac{1}{2}$, with values of the form $-\tfrac{1}{2} \pm j\omega$. The result is shown in Fig. 7-5.

What does this tell us about the response of the closed loop system? For small k, much less than $\tfrac{1}{4}$, the system will be slow to settle due to a pole near the origin. At $k = \tfrac{1}{4}$, the system will be critically damped with time constants of 2 seconds. As k increases further, the response starts to exhibit a damped sinusoid; increasing the gain increases the frequency of the sinusoid, but the envelope remains unchanged as an exponential with a decay time constant of 2 seconds.

Although we have talked of poles and zeros, the only systems we have looked at so far have only had poles. Let us consider a system with one of each:

$$G = \frac{s + 1}{s + 2}$$

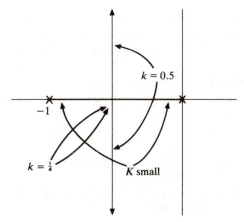

Figure 7-5 Roots of $s^2 + s + k = 0$.

The closed loop gain is

$$\frac{k(s + 1)}{s + 2 + k(s + 1)} = \frac{k(s + 1)}{(1 + k)s + (2 + k)}$$

Clearly the closed loop system has a zero in exactly the same place as the open loop. This is not surprising. If G gives zero output for some particular value of complex input frequency, then no amount of feedback will make it change its mind.

There is just one pole, at a value $-(2 + k)/(1 + k)$. If k is small, the pole is near its open loop position. As k increases, the pole moves along the axis until at very large k it approaches the position of the zero at -1.

Now let us try two poles and one zero:

$$G(s) = \frac{s + 1}{s(s + 2)}$$

This time the closed loop gain is

$$\frac{k(1 + s)}{s^2 + 2s + k(1 + s)}$$

so the poles are at the roots of

$$s^2 + (2 + k)s + k = 0$$

that is

$$s = -\frac{2 + k}{2} \pm \sqrt{\left(1 + \frac{k}{2}\right)^2 - k}$$

$$= -1 - \frac{k}{2} \pm \sqrt{1 + \frac{k^2}{4}}$$

Both roots are real. For small k the roots are near -2 and 0, where we would expect. As k tends to infinity, the square root tends to the value $k/2$, and we find one root near -1 while the other is dashing off along the negative real axis.

What have we learned from these three simple examples?
1. The closed loop system has the same zeros as the open loop.
2. As k tends to infinity, a pole approaches the zero.
3. If we have two poles and one zero, the 'spare' pole heads off along the negative real axis.
4. If we have two poles and no zeros, so that both are 'spare', then they head off north and south in the imaginary directions.

Can these properties be generalized to a greater number of poles and zeros, and what other techniques can be discovered?

7-4 POLES AND POLYNOMIALS

We can represent the gain as a ratio of two polynomials, as in expression (7-1), or more succinctly in the form

$$G(s) = \frac{Z(s)}{P(s)}$$

allowing the closed loop gain to be written as

$$\frac{kG}{1 + kG} = \frac{kZ(s)}{P(s) + kZ(s)}$$

To plot the progress of the poles, we must look at the roots of

$$P(s) + kZ(s) = 0 \tag{7-2}$$

Without losing generality we can absorb the constant a of G into the value of k. Now when we multiply out $P(s)$ and $Z(s)$ the coefficient of the highest power of s in each of them is unity.

If there are n poles and m zeros, $P(s)$ and $Z(s)$ are polynomials of degree n and m respectively and will appear in the form:

$$Z(s) \equiv s^m + a_{m-1}s^{m-1} + \cdots + a_0 = 0$$

and

$$P(s) \equiv s^n + b_{n-1}s^{n-1} + \cdots + b_0 = 0$$

We can draw a number of different conclusions according to whether $m = n, m = n - 1$ or $m < n - 1$. We will rule out the possibility of having more zeros than poles, since that would imply that the gain increased indefinitely with increasing frequency.

1. If there are equal numbers of poles and zeros, the polynomial (7-2) can be expanded term by term as

$$(1 + k)s^n + (b_{n-1} + ka_{n-1})s^{n-1} + \cdots + (b_0 + ka_0) = 0$$

As k becomes very large, the a term will dominate the b contribution in every one of the coefficients. Indeed, the equation will have the same roots as

$$s^n + \left(\frac{b_{n-1}}{1 + k} + a_{n-1}\frac{k}{1 + k}\right)s^{n-1} + \cdots + \left(\frac{b_0}{1 + k} + a_0\frac{k}{1 + k}\right) = 0$$

Clearly the final resting place of the n poles will be at the location of the n zeros.

2. If $m = n - 1$, then the leading term is unchanged by k, giving a polynomial

$$s^n + (b_{n-1} + ka_{n-1})s^{n-1} + \cdots + (b_0 + ka_0) = 0$$

If we divide through by k, we are left with an equation that for large k starts to look like

$$\frac{s^n}{k} + Z(s) = 0$$

Now $n - 1$ of the closed loop poles are homing in on the roots of Z, but we cannot just lose the last one. To track it down, we write out another approximation to the polynomial:

$$\left(\frac{s}{k} + 1\right)Z(s) = 0$$

i.e. the missing pole has raced off along the negative real axis with a value $-k$.

One pole thus makes for $-\infty$ while the others settle down at the values of the zeros.

3. When m is less than $n - 1$, the analysis becomes much more interesting. Before tangling with the algebra, however, let us examine the polynomials P and Z more closely.

$Z(s)$ and $P(s)$ appear in the form

$$s^m + a_{m-1}s^{m-1} + \cdots + a_0$$

and

$$s^n + b_{n-1}s^{n-1} + \cdots + b_0$$

In Eq. (7-1), these same polynomials were written in terms of their roots, as a product of terms $(s - p_i)$ or $(s - z_i)$. When we multiply Z out, we see at once that

$$a_{m-1} = -\sum z_i$$

the negative of the sum of all the zeros. Similarly,

$$b_{n-1} = -\sum p_i$$

the negative of the sum of all the poles. How will this help us to factorize the polynomial

$$P(s) + kZ(s)$$

into its roots? Let $n = m + r$, where r is at least 2.

We can attempt a long division of $P(s)$ by $Z(s)$. The first term of the ratio is s^r, so we subtract $s^r Z(s)$ from $P(s)$ to leave

$$(b_{n-1} - a_{m-1})s^{n-1} + (b_{n-2} - a_{m-2})s^{n-2} + \cdots$$

The next term of the ratio is $(b_{n-1} - a_{m-1})s^{r-1}$, so we can continue the long division process until we come to a remainder polynomial of a lower order than $Z(s)$. If we call this $R(s)$, then we can write

$$P(s) = [s^r + (b_{n-1} - a_{m-1})s^{r-1} + \cdots]Z(s) + R(s)$$

so

$$P(s) + kZ(s) = [s^r + (b_{n-1} - a_{m-1})s^{r-1} + \cdots + k]Z(s) + R(s)$$

As k tends to infinity, the coefficients of R become swamped by the coefficients of $kZ(s)$, and the polynomial approximates to

$$[s^r + (b_{n-1} - a_{m-1})s^{r-1} + \cdots + k]Z(s)$$

When we equate this to zero it is no surprise to find a set of roots tending to the zeros; it is the remaining roots that are of interest.

The polynomial

$$[s^r + (b_{n-1} - a_{m-1})s^{r-1} + \cdots + k] \tag{7-3}$$

has a leading term s^r and a large constant term k. To a very rough approximation, the roots will look like the rth roots of $-k$. If $r = 2$, they will lie in the $+j$ and $-j$ directions, heading north and south.

If $r = 3$, these roots will go in the directions of the cube roots of -1; one will go due west along the negative real axis, while the others will go roughly north-east and south-east at $120°$ angles. These directions are inevitably unstable for sufficiently large k, so we see why our troubles really start when there are three excess poles.

We have a rough idea of the directions of the asymptotes, but can we define them any more clearly? Suppose that they intersect at $s = c$. We might be able to approximate the roots as solutions of

$$(s - c)^r + k = 0.$$

If we multiply out the first term we obtain

$$s^r - rcs^{r-1} + \cdots + (-c)^r + k = 0$$

If we equate the coefficients of s^{r-1} in this and expression (7-3), we obtain

$$-rc = b_{n-1} - a_{m-1}$$

Remembering that $r = n - m$ and that $-b_{n-1}$ and $-a_{m-1}$ are the sums of the values of the poles and the zeros respectively, we have

$$c = \frac{1}{n - m} \left(\sum p_i - \sum x_i \right)$$

Give each pole a weight of plus one unit and each zero a weight of minus one unit and the intersection of the asymptotes will be found at the centre of gravity.

Exercise 7-4-1 Show that the root locus of the system $1/[s(s + 1)^2]$ has three asymptotes which intersect at $s = -\frac{2}{3}$. Make a very rough sketch.

Exercise 7-4-2 Add a zero at $s = -2$, so that the system has transfer function $(s + 2)/[s(s + 1)^2]$. What and where are the asymptotes now?

7-5 BREAKAWAY POINTS AND THE REAL AXIS

Let us once again think of the root locus as the set of values of s for which G is real and negative. If we have just a pair of complex poles, then every point of the real s axis will give a value of G that is real and positive. The phase angle contributions of such a pair of poles will cancel out on the real axis.

If we have just one pole, say at $s = -1$, then at all points on the real axis to the left of $s = -1$, G will be negative, while to the right G will be positive. If we add a second pole on the axis, then to the left of both poles G will be positive again, while between them it is negative. The same goes for adding a real zero instead. By adding up phase angles of vectors joining each pole or zero to s, we see that G is negative at points on the real axis where there are an odd number of real poles or zeros to the right of s.

As an example, consider again the system $1/[s(s + 1)]$, with poles at $s = 0$ and $s = -1$. To the left of both poles, there are two poles to the right and so G is positive. This part of the axis is not part of the plot. Between the poles G has one pole to the right, so is negative. To the right of both poles G is positive once again.

As the gain increases, the poles move along the axis towards each other. At $s = -\frac{1}{2}$ they join forces, then break away north and south along their asymptotes. This has already been illustrated in Fig. 7-5.

Take as another example the system of Exercise 7-4-1. There are now two poles at $s = -1$ and one at $s = 0$. To the left of all three poles, at $s = -5$ for example, G will be negative so this part of the real axis forms part of the root locus plot. Between the poles, there will be two to the left and one to the right. We can ignore those to the left, but the single pole to the right tells us that G is negative and this is also part of the plot. To the right of $s = 0$ we run out of poles, so the plot stops there.

As k starts to increase, the poles at $s = -1$ part company and one moves off in the negative direction. The other moves towards the $s = 0$ pole, which in turn is moving to the left. These poles join forces somewhere between $s = -1$ and $s = 0$ and then break away to approach their asymptotes. Can we calculate the value of s at which they will break apart?

Remember the properties of the mapping function. If the derivative of $G(s)$ is written as $G'(s)$, then a small change δs will result in a change of gain $\delta G = G'(s)\, \delta s$. Looking forward and backward along the line $G = $ real, δG is also a small real increment, positive or negative. The inverse mapping of this line may bend, but how can it fork? Clearly something has gone wrong with $G'(s)$ at the point in question.

If we consider a few more terms of the power series expansion for δG, we see that

$$\delta G = G'(s)\, \delta s + \frac{G''(s)(\delta s)^2}{2} + \frac{G'''(s)(\delta s)^3}{3!} + \cdots$$

Now all becomes clear. If $G'(s)$ becomes zero, then we are left with δG being proportional to δs^2, that is δs is proportional to the square root of $+\delta G$ or $-\delta G$; it has four possible values and directions, at 90° intervals.

If both $G'(s)$ and $G''(s)$ are zero, then δs behaves as the cube root of δG and we have a junction of six paths, now at 60° intervals.

Of course, $G'(s)$ could vanish 'off the beaten track' without resulting in a breakaway point.

Let us take up the story of our example again. If

$$G(s) = \frac{1}{s(s + 1)^2}$$

then

$$\frac{\mathrm{d}}{\mathrm{d}s} G(s) = -\frac{(s + 1) + 2s}{s^2(s + 1)^3}$$

This is zero at $s = -\frac{1}{3}$, which lies on the locus. We can finally gather all the fragments of information to sketch the root locus as shown in Fig. 7-6.

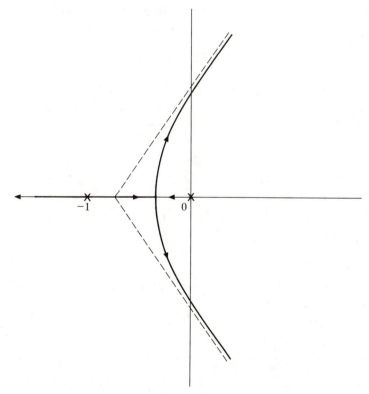

Figure 7-6 Root locus of $1/s(s + 1)^2$.

One last point concerns the evaluation of the direction of the root locus at any point. Since $\delta G = G'(s)\,\delta s$ and since δG is real we can work out the direction of δs by differentiating $G(s)$ and taking the argument of its inverse. This is really of interest when we wish to find out the direction in which the locus leaves off-axis poles or joins off-axis zeros.

To sum up:

1. If there are n poles and m zeros, the $n - m$ 'extra' poles make off towards infinity along asymptotes in the directions of the $(n - m)$th roots of -1.
2. These asymptotes all meet at a point on the real axis. To locate this point, assign a weight of $+1$ to each pole and -1 to each zero, and find the centre of gravity. (Or numerically compute the sum of the poles, subtract the sum of the zeros and divide the result by $n - m$.)
3. Those parts of the real s axis that have an odd number of real poles or zeros to the right will form part of the root locus. (The remainder will correspond to values of G that are real and positive.)
4. At any junctions (breakaway points) the derivative of $G(s)$ will be zero (but not all zeros of $G'(s)$ will necessarily lie on the locus).
5. The direction of the locus can be calculated at any value of s by evaluating $1/G'(s)$. This will give the departure angles from off-axis poles or zeros.

7-6 COMPENSATORS AND OTHER EXAMPLES

We have so far described the root locus as though it were only applicable to unity feedback. Suppose that we use some controller dynamics, either at the input to the system or in the feedback loop (see Fig. 7-7).

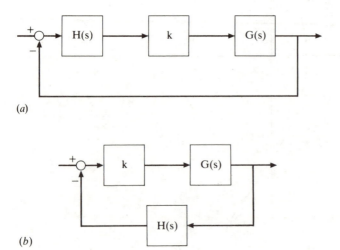

(a)

(b)

Figure 7-7 Loops with dynamic compensation.

The closed loop gains in these two cases are

$$\frac{kH(s)G(s)}{1 + kH(s)G(s)}$$

and

$$\frac{kG(s)}{1 + kH(s)G(s)}$$

respectively. Although the gains may be different, the polynomial we look at to assess the closed loop poles will be exactly the same in each case. We just add the compensator's poles and zeros into the pot along with those of the open loop system.

Exercise 7-6-1   *An undamped motor has response $1/s^2$. With a gain k in front of the motor and unity feedback around the loop, sketch the root locus. Does it look encouraging?*

Exercise 7-6-2   *Now apply phase advance, by inserting $H(s) = (s + 1)/(s + 3)$ in front of the motor. Does the root locus look any more hopeful?*

Exercise 7-6-3   *Change the phase advance to that of Exercise 5-7-1, $H(s) = (3s + 1)/(s + 3)$. Are there any breakaway points?*

Let us work out these exercises here. The system $1/s^2$ has two poles at the origin. There are two excess poles, so there are two asymptotes in the positive and negative imaginary directions. The asymptotes pass through the 'centre of gravity', i.e. through $s = 0$. No part of the real axis can form part of the plot, since both poles are encountered together.

We deduce that the poles split immediately and make off up and down the imaginary axis. For any value of negative feedback, the result will be a pair of pure imaginary poles representing simple harmonic motion.

Now let us add phase advance in the feedback loop, with an extra pole at $s = -3$ and a zero at $s = -1$. There are still two excess poles, so the asymptotes are still parallel to the imaginary axis. However, they will no longer pass through the origin.

To find their intersection, take moments of the poles and zero. We have contribution 0 from the poles at the origin, -3 from the other pole and $+1$ from the zero. The total, -2, must be divided by the number of excess poles to find the intersection: at $s = -1$.

How much of the axis forms part of the plot? Between the pole at -3 and the zero, there is one real zero plus two poles to the right of s—an odd total. To the left of the single pole and to the right of the zero the total is even.

Are there any breakaway points?

$$G(s) = \frac{s + 1}{s^2(s + 3)}$$

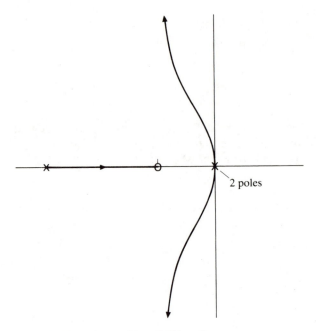

Figure 7-8 Root locus of $(s + 1)/s^2(s + 3)$.

Differentiating, we find that

$$G'(s) = -2\,\frac{s^2 + 3s + 3}{s^3(s + 3)^2}$$

This becomes zero at $s = (-3 \pm \sqrt{3}\,j)/2$; these points in a preliminary sketch look unlikely to lie on the locus.

Putting all these deductions together, we arrive at a sketch as shown in Fig. 7-8. The system is safe from instability. For large values of feedback gain, the resonance poles resemble those of a system with added velocity feedback (see Fig. 7-5).

Now let us look at Exercise 7-6-3. It looks very similar in format, except that the phase advance is much more pronounced. The high frequency gain of the phase advance term is in fact nine times its low frequency value.

We have two poles at $s = 0$ and one at $s = -3$, as before. The zero is now at $s = -\frac{1}{3}$. For the position of the asymptote pair, we have a moment -3 from the lone pole and $+\frac{1}{3}$ from the zero. The asymptotes thus cut the real axis at half this total, at $-\frac{4}{3}$.

As before, the only part of the real axis to form part of the plot is that joining the singleton pole to the zero. It looks as though the plot may be very similar to the last.

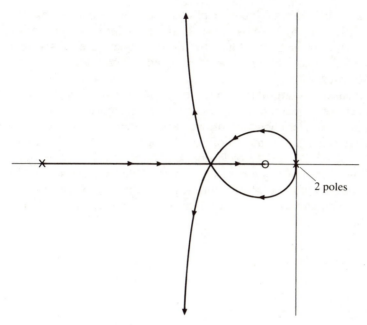

Figure 7-9 Root locus of $(3s + 1)/s^2(s + 3)$.

Let us just look at the breakaway condition, though:

$$3G(s) = \frac{3s + 1}{s^2(s + 3)} \tag{7-4}$$

from which we find that

$$3G'(s) = \frac{-6(s^2 + 2s + 6)}{s^3(s + 3)^2}$$

Now we have not just one root at $s = -1$ but two. Both $G'(s)$ and $G''(s)$ vanish here, at a point well and truly lying on part of the plot. We have a six-way branch, and the resulting picture is shown in Fig. 7-9.

At the branch point, at $s = -1$, the gain is found by substituting into Eq. (7-4), to be $\frac{1}{3}$. A gain value $k = 3$ will therefore put all three poles at a value of $s = -1$, and the response will be very well damped indeed.

7-7 CONCLUSION

The root locus gives a remarkable insight into the selection of the value of a feedback parameter. It enables phase advance and other compensators to be considered in an educated way. It can be plotted automatically by computer, or with only a little effort by hand by the application of relatively simple rules.

A lot of effort has been devoted here to this technique, since we will later find it to be effective for the analysis of sampled systems too. It has its restrictions, however.

The root locus in its natural form only considers the variation of a single parameter. When we have multiple inputs and outputs, although we can still consider a single characteristic equation we have a great variety of possible feedback arrangements. The same set of closed loop poles can sometimes be achieved with an infinite variety of feedback parameters, and some other basis must be used for making a choice. With multiple feedback paths, the zeros are no longer a fixture so that individual output responses can be tailored. Other considerations can be non-linear ones of drive saturation or energy limitation.

MORE MATRIX ALGEBRA. CONVOLUTION

P8-1 INTRODUCTION

Two separate topics have to be covered before the next chapter. The first involves some more matrix algebra and concentrates on generalizing the process of inverting a matrix. Some attention will also be given to the 'characteristic equation'.

The second topic is that of 'convolution', another trip into the world of infinite integrals in which integral transforms abound. There is a computer program or two to try to give a practical 'feel' to the material, which exchanges the concept of the transfer function for that of the impulse response.

P8-2 THE INVERSE OF A MATRIX

Back in Sec. 3-4 we came across a linear relationship between two vectors,

$$\mathbf{w} = T\mathbf{x}$$

and carelessly deduced that $\mathbf{x}$ could be expressed in terms of $\mathbf{w}$ by

$$\mathbf{x} = T^{-1}\mathbf{w}$$

If such a thing exists, T^{-1} must be the 'inverse' of the matrix T, and must satisfy

$$T^{-1}T = I$$

where I is the unit matrix of the same size. What are the conditions for being able to find an inverse for a given matrix and how can it be computed?

In Sec. P3-4 we considered a 2 by 2 example. By the straightforward expedient of solving the two simultaneous equations it was shown that the inverse of

$$\begin{bmatrix} a & b \\ c & d \end{bmatrix}$$

was

$$\frac{1}{ad - bc} \begin{bmatrix} d & -b \\ -c & a \end{bmatrix}$$

How does this extend to larger matrices?

The answer lies in the concept of the *determinant* of a matrix. This is just a single number, whatever the size of the matrix. In the example above, the determinant equals $(ad - bc)$. Its value to the applied mathematician or control engineer lies in the fact that it is straightforward to compute and that numerous useful results can be drawn from it, but it is perhaps more satisfying to check out the pure mathematical background.

While discussing generalities, we will keep an eye on the particular 3 by 3 case:

$$\begin{bmatrix} a & b & c \\ d & e & f \\ g & h & i \end{bmatrix}$$

Now given the numbers $(1, 2, 3, \ldots, n)$, we can form a *permutation* by repeatedly swapping pairs of numbers; thus $(1, 2, 3)$ can become $(1, 3, 2)$ and then $(3, 1, 2)$. We can then use this permutation to pick one term from each row of the matrix, so that no column is used twice. The permutation $(3, 1, 2)$ would select terms c, d and h.

To each permutation corresponds a *signature*, a value $+1$ or -1, easily evaluated by counting whether the permutation has required an even or an odd number of swaps—an even number gives $+1$ while an odd number gives -1. The permutation $(3, 1, 2)$ required two swaps, so its signature is positive.

Having used the permutation to select the terms c, d and h, we multiply them together, and then multiply by the signature to obtain $+cdh$.

The mathematical definition of the determinant of a matrix is the sum of all such terms over all possible permutations. In the 3 by 3 example, we have permutations

$$(1, 2, 3) \quad (2, 3, 1) \quad (3, 1, 2) \quad (1, 3, 2) \quad (3, 2, 1) \quad \text{and} \quad (2, 1, 3)$$

giving corresponding terms in the determinant

$$+aei \quad +bfg \quad +cdh \quad -afh \quad -ceg \quad -bdi$$

Six terms are not too much trouble here, but beware! There are over half a million permutations of ten numbers! We must try to establish some system.

If we collect top-row terms and take out factors, we arrive at

$$a(ei - fh) + b(fg - di) + c(dh - eg)$$

But now the terms in brackets look suspiciously like determinants of 2 by 2 matrices, and this is of course just what they are. The first such term, $(ei - fh)$, is the determinant of the matrix obtained by rubbing out the top row and first column of the matrix we started with. Likewise the third term is obtained by rubbing out the row and column through c. The middle term is just slightly different. We rub out the row and column through b, but then must multiply the resulting determinant by -1.

From our original matrix, we can compute a 'matrix of cofactors', which we write as

$$\begin{bmatrix} A & B & C \\ D & E & F \\ G & H & I \end{bmatrix}$$

where D, for example, is obtained by taking the determinant of the 2 by 2 matrix left when the row and column through element d are rubbed out and then multiplying in this case by -1. The sign for multiplication is given by the corresponding element of

$$\begin{bmatrix} + & - & + \\ - & + & - \\ + & - & + \end{bmatrix}$$

Now we find that the determinant, Δ, is given by $aA + bB + cC$. If we instead gather together terms in the second row, we discover that the determinant is also given by $dD + eE + fF$, and the same for the third row. We could equally have gathered together terms in the first column to give the value of the determinant as $aA + dD + gG$. We can list these results as:

1. The value of the determinant

$$\Delta = aA + bB + cC = dD + eE + fF = gG + hH + iI$$

$$= aA + dD + gG = bB + eE + hH = cC + fF + iI$$

We can evaluate the determinant by 'expanding' by any of the rows or columns, to obtain one of the expressions listed above.

What happens if we swap two columns of the matrix? Each of the product terms will occur as before, but each permutation will start off with a swap already made. Its signature will therefore be the negative of the previous value associated with that term. The total result will be to give a determinant that is the previous value multiplied by -1. Swapping two rows will have the same result.

2. Swapping two rows, or swapping two columns, will negate the value of the determinant.

Suppose that two columns are the same. Swapping them will negate the determinant, but the matrix and the determinant are clearly unchanged. The only possibility is that the determinant is zero. The same goes for repeated rows.

3. If two rows are the same, or if two columns are the same, then the determinant is zero.

Suppose that we double the values of the top row elements. None of the cofactors of the top row will be affected. By expanding by the top row, we then see that the value of the determinant is also doubled.

4. If all elements of a row or of a column are multiplied by a constant, the determinant will be multiplied by the same constant.

If we add p, q and r respectively to the top row, how is the value of the determinant changed? We can expand by the top row to show

$$\det \begin{bmatrix} a+p & b+q & c+r \\ d & e & f \\ g & h & i \end{bmatrix}$$

$$= (a+p)A + (b+q)B + (c+r)C$$

$$= aA + bB + cC + pA + qB + rC$$

$$= \det \begin{bmatrix} a & b & c \\ d & e & f \\ g & h & i \end{bmatrix} + \det \begin{bmatrix} p & q & r \\ d & e & f \\ g & h & i \end{bmatrix}$$

Now what happens if (p, q, r) is a multiple of one of the other rows? From results 3 and 4 we know that the second of the two determinants just above must be zero, and so adding (p, q, r) to the top row leaves the total determinant unchanged. We can generalize this as follows.

5. If to one of the rows of the matrix we add a multiple of another row, or to one of the columns we add a multiple of another column, then the value of the determinant is unchanged.

How does all this help us with inverting a matrix? Consider the product of our matrix with the transpose of its matrix of cofactors:

$$\begin{bmatrix} a & b & c \\ d & e & f \\ g & h & i \end{bmatrix} \begin{bmatrix} A & D & G \\ B & E & H \\ C & F & I \end{bmatrix}$$

$$= \begin{bmatrix} aA+bB+cC & aD+bE+cF & aG+bH+cI \\ dA+eB+fC & dD+eE+fF & dG+eH+fI \\ gA+hB+iC & gD+hE+iF & gG+hH+iI \end{bmatrix}$$

Look closely at these terms. The top left gives us the determinant, as do the centre and the bottom right terms. The second term in the left-hand column is $dA + eB + fC$. This is just the result we would get if we replaced the top row of the matrix by repeating the second row and then took its determinant; the top two rows are the same, so the value is zero. The same is true for every one of the 'mismatched' terms, so the only non-zero values will be on the diagonal. We see that the product above gives us

$$\begin{bmatrix} \Delta & 0 & 0 \\ 0 & \Delta & 0 \\ 0 & 0 & \Delta \end{bmatrix}$$

If we divide each of the terms of the transposed matrix of cofactors by the determinant of the matrix, we will thus have arrived at the inverse. We can sum up as follows.

The inverse of a matrix can be found by computing the matrix of cofactors, transposing it and dividing term by term by the determinant of the matrix. It will be both a *left inverse* and a *right inverse*, i.e. the product with the original matrix taken in either order will give the unit matrix.

Although this is all mathematically sound, it is not the quickest way of inverting a large matrix, as we will see.

P8-3 COMPUTING THE MATRIX INVERSE

By the inverse of the matrix T, we mean a matrix which when multiplied by T will give the unit matrix. Suppose we can find a collection of elementary matrices, P, Q, R, S and so on, so that multiplying T by each in turn will 'break it down' towards the unit matrix. Then their product will give the inverse, if taken strictly in the correct order.

Consider the product

$$\begin{bmatrix} 1 & k & 0 \\ 0 & 1 & 0 \\ 0 & 0 & 1 \end{bmatrix} \begin{bmatrix} a & b & c \\ d & e & f \\ g & h & i \end{bmatrix}$$

giving

$$\begin{bmatrix} a + kd & b + ke & c + kf \\ d & e & f \\ g & h & i \end{bmatrix}$$

Premultiplying by the first matrix has had the effect of adding k times the second row of T to the first row. Similar elementary matrices can be found which will multiply a row by a constant or exchange rows. Note that we must stick to row operations; we must not mix them with column operations. (We could, of course, start from scratch with column operations alone, which would be equivalent to postmultiplication by elementary matrices.)

Now suppose that we subtract d/a times the top row of T from the second row. The d term will be replaced by zero. Subtract g/a of the top row from the third, and we have another zero. Each of these is equivalent to premultiplying by a matrix, and provided we keep track of the operations by performing self-same operations on the unit matrix we end up by constructing their product.

We can swiftly reduce the matrix to *upper triangular form*, and with a little more effort whittle it down to the unit matrix. Let us take an example:

Test matrix

$$\begin{bmatrix} 1 & 1 & 3 \\ 2 & 7 & 9 \\ 1 & 5 & 6 \end{bmatrix}$$

Unit matrix

$$\begin{bmatrix} 1 & 0 & 0 \\ 0 & 1 & 0 \\ 0 & 0 & 1 \end{bmatrix}$$

twice top row from middle:

$$\begin{bmatrix} 1 & 1 & 3 \\ 0 & 5 & 3 \\ 1 & 5 & 6 \end{bmatrix}$$

$$\begin{bmatrix} 1 & 0 & 0 \\ -2 & 1 & 0 \\ 0 & 0 & 1 \end{bmatrix}$$

top row from bottom:

$$\begin{bmatrix} 1 & 1 & 3 \\ 0 & 5 & 3 \\ 0 & 4 & 3 \end{bmatrix}$$

$$\begin{bmatrix} 1 & 0 & 0 \\ -2 & 1 & 0 \\ -1 & 0 & 1 \end{bmatrix}$$

bottom row times 5 minus 4 times middle row:

$$\begin{bmatrix} 1 & 1 & 3 \\ 0 & 5 & 3 \\ 0 & 0 & 3 \end{bmatrix}$$

$$\begin{bmatrix} 1 & 0 & 0 \\ -2 & 1 & 0 \\ 3 & -4 & 5 \end{bmatrix}$$

Now mop up the rest of the off-diagonal terms: bottom row from top row and bottom row from middle (take care!):

$$\begin{bmatrix} 1 & 1 & 0 \\ 0 & 5 & 0 \\ 0 & 0 & 3 \end{bmatrix}$$

$$\begin{bmatrix} -2 & 4 & -5 \\ -5 & 5 & -5 \\ 3 & -4 & 5 \end{bmatrix}$$

top row times 5 minus second row:

$$\begin{bmatrix} 5 & 0 & 0 \\ 0 & 5 & 0 \\ 0 & 0 & 3 \end{bmatrix}$$

$$\begin{bmatrix} -5 & 15 & -20 \\ -5 & 5 & -5 \\ 3 & -4 & 5 \end{bmatrix}$$

and finally divide top row and middle row by 5, bottom row by 3:

$$\begin{bmatrix} 1 & 0 & 0 \\ 0 & 1 & 0 \\ 0 & 0 & 1 \end{bmatrix} \qquad \begin{bmatrix} -1 & 3 & -4 \\ -1 & 1 & -1 \\ 1 & -4/3 & 5/3 \end{bmatrix}$$

Exercise P8-3-1 *Multiply the matrix on the right by the original matrix to verify that it is indeed the inverse.*

The method involves a large number of petty operations—just the stuff that computers thrive on. The computer does not grumble when the numbers become awkward, either. With very little additional elaboration, the method can be transformed into a computer program requiring a number of operations only proportional to the square of the matrix size.

P8-4 THE CHARACTERISTIC POLYNOMIAL

Before we leave the subject of matrices and their determinants, let us look at the inverse of the matrix $(sI - A)$, where A is some square matrix and I is the unit matrix of the same size.

If we invert $sI - A$ in the 'classic' way, by computing a matrix of cofactors, then these cofactors will be polynomials in s. To obtain the inverse, we must divide each cofactor by the determinant of $(sI - A)$, and we see that this must also be a polynomial in s. If A is n by n, then the polynomial has degree n and is called the *characteristic polynomial* of the matrix A. When equated to zero, its roots will be of great future interest to us.

Now if we look at $sI - A$, we will see terms of the form $(s - a_{ii})$ along the main diagonal, with terms $(-a_{ij})$ elsewhere. Only from one term in the expansion of the determinant, that from the main diagonal, will we find n s's multiplied together. Furthermore, any other permutation must give at least two less $(s - a_{ii})$ products, so the terms in s^{n-1} can also only come from the diagonal. We can deduce that the sum of the roots of the polynomial will equal the 'trace' of the matrix, the sum of the diagonal coefficients of A.

Exercise 8-4-1 *By considering the constant term, show that the product of the roots must be equal to the determinant of A.*

P8-5 CONVOLUTION

Now we have a complete change of scene, as we look again at the world of multiple integrals and transforms.

We saw that if we applied a signal with Laplace transform $U(s)$ to a system $H(s)$, then the transform of the output, $Y(s)$, satisfied

$$Y(s) = H(s)U(s)$$

If we could find the input whose transform is just 1 and apply it to the system, then the output would be just $H(s)$. This would be the transform of a very interesting time function peculiar to the system, which we could call $h(t)$.

The transform of the unit step, $y = 1$ for $t > 0$, is easily found by integrating $[1 . \exp(-st)]$ from zero to infinity. It has value $1/s$. The Laplace form of a perfect integrator is also $1/s$. If we apply our unknown test signal to an integrator, therefore, we will get an output that is a unit step at $t = 0$.

Since the integrator must give a step change, the test signal must be infinitely tall. However, its integral is just unity. Imagine it as the limit of a rectangular pulse of unit area, as the rectangle becomes vanishingly narrow (see Fig. P8-1). This signal, whose transform is 1, is the 'unit impulse'.

If we apply such an impulse to a system $H(s)$, the output $h(t)$ will have the transform $H(s)$. $h(t)$ is termed the *impulse response* of the system.

Now if we apply an input of size $u(0)\, \delta t$ at $t = 0$, the output at time 5 will be $h(5)u(0)\, \delta t$. If, in addition, we apply an input $u(1)\, \delta t$ at $t = 1$, the output at $t = 5$ will be the sum of a five-second-old response, as before, plus a new four-second-old response:

$$h(5)u(0)\, \delta t + h(4)u(1)\, \delta t$$

A further impulse at time t will add a term

$$h(5 - t)u(t)\, \delta t$$

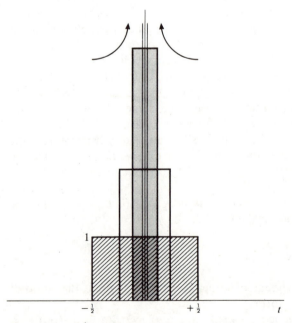

Figure P8-1 The unit impulse has unit area. It is the limit as a rectangular pulse is made vanishingly narrow.

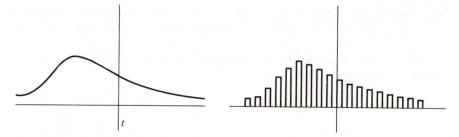

The input can be broken into a train of impulses.

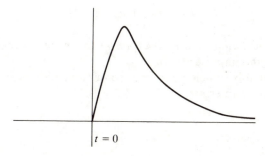

To each impulse the system will respond with its
impulse response.

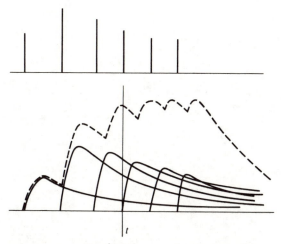

The response is the sum of the individual impulse responses.
The value at time t depends only on earlier input impulses.

Figure P8-2 Superposition of impulse responses.

Any impulse arriving after $t = 5$ will be too late to affect the output, unless the system is equipped with a crystal ball for future gazing. This is illustrated in Fig. P8-2.

We can generalize this for a large number of input impulses $u(t)\,\delta t$, to see that the output at time T is now

$$y(T) = \sum_{t=0}^{T} h(T - t)u(t)\,\delta t$$

If we now consider $u(t)$ to be a continuous function, we can break it into an infinite succession of infinitesimal impulses and the summation becomes an integral:

$$y(T) = \int_{t=0}^{T} h(T - t)u(t)\,\delta t$$

which is the *convolution integral*. It expresses the output as the convolution of the input with the impulse response.

This is probably sufficient for an introduction; matters are carried further in the following chapter.

EIGHT

LINKING THE TIME AND FREQUENCY DOMAINS

8-1 INTRODUCTION

The state-space description of a system can come close to describing its ingredients; state equations can be built from the physical relationships between velocities and positions, levels, flows and any other identifiable quantities. The frequency domain description is more concerned with responses that can be measured at the outputs when signals are applied to the inputs, regardless of what may be contained in the 'black box'. We have looked at aspects of both techniques in isolation; now let us tie the two areas together.

8-2 STATE-SPACE AND TRANSFER FUNCTIONS

In Chapter 3, when considering a position control system, we dithered between state equations and second-order differential equations and switched fairly easily between the two. A second-order differential equation, such as

$$\ddot{y} + 5\dot{y} + 6y = 6u$$

expresses a clear relationship between the single input, the position demand and the output position of the motor. It would be the work of a moment to sprinkle a flavouring of Laplace onto this equation and start to apply any of the methods of the root locus.

A set of state equations describes exactly the same system, but in formalized terms, tying position and velocity into simultaneous first-order equations, and by expressing the output as a mixture of these state variables we obtain

$$\begin{bmatrix} \dot{x}_1 \\ \dot{x}_2 \end{bmatrix} = \begin{bmatrix} 0 & 1 \\ -6 & -5 \end{bmatrix} \begin{bmatrix} x_1 \\ x_2 \end{bmatrix} + \begin{bmatrix} 0 \\ 6 \end{bmatrix} u$$

$$y = \begin{bmatrix} 1 & 0 \end{bmatrix} \begin{bmatrix} x_1 \\ x_2 \end{bmatrix}$$

(8-1)

This last equation does no more than select the motor position as our output, yet involves multiplying a matrix and a vector. This formal method is most appealing when we have a computer at our fingertips; a computer will perform a vast number of matrix multiplications in the time it takes for us to be ingenious.

With a single input and a single output, we find a single transfer function linking the two, just as we would expect. If we have two inputs and one output, and if the system is linear, then the output will be made up of the sum of the effects of the two inputs. We can find two transfer functions, one associated with each input, allowing the output to be expressed as

$$Y(s) = G_1(s)U_1(s) + G_2(s)U_2(s)$$

In the same way, one input and two outputs will also have two transfer functions.

When we move up to two inputs and two outputs, we discover four transfer functions. Each output is linked to each of the two inputs, and this can be represented most neatly in matrix form:

$$\begin{bmatrix} Y_1(s) \\ Y_2(s) \end{bmatrix} = \begin{bmatrix} G_{11}(s) & G_{12}(s) \\ G_{21}(s) & G_{22}(s) \end{bmatrix} \begin{bmatrix} U_1(s) \\ U_2(s) \end{bmatrix}$$

where each of the G's will probably be a ratio of two polynomials.

To gain the greatest advantage from the computer, we would like to find routine ways of moving between state-space and transfer function forms.

8-3 DERIVING THE TRANSFER FUNCTION MATRIX

We wish to find a transfer function representation for a system given in state-space form. We start with

$$\dot{x} = A\mathbf{x} + B\mathbf{u}$$
$$y = C\mathbf{x}$$

(8-2)

and

Since we are looking for a transfer function, it seems a good idea to take the Laplace transform of $\mathbf{x}(t)$ and deal with $\mathbf{X}(s)$. In Sec. P6-5 we saw that

$$\mathscr{L}\{\dot{\mathbf{x}}(t)\} = s\mathbf{X}(s) - \mathbf{x}(0)$$

so

$$s\mathbf{X}(s) - \mathbf{x}(0) = A\mathbf{X}(s) + B\mathbf{U}(s)$$

that is

$$s\mathbf{X}(s) - A\mathbf{X}(s) = \mathbf{x}(0) + B\mathbf{U}(s)$$

Now we can mix the two terms on the left together more easily if we write

$$s\mathbf{X}(s)$$

as

$$sI\mathbf{X}(s)$$

where I is the unit matrix. This enables us to take out $\mathbf{X}(s)$ as a factor, to obtain

$$(sI - A)\mathbf{X}(s) = \mathbf{x}(0) + B\mathbf{U}(s)$$

All that remains is to find the inverse of the matrix $sI - A$ and we can multiply both sides by it to obtain a clear $\mathbf{X}(s)$ on the left. Remember that the order of multiplication is important. The inverse must be applied in front of each side of the equation to give

$$\mathbf{X}(s) = (sI - A)^{-1}[\mathbf{x}(0) + B\mathbf{U}(s)]$$

Now we also have

$$\mathbf{Y}(s) = C\mathbf{X}(s)$$

so the outputs are related to the inputs by

$$\mathbf{Y}(s) = C(sI - A)^{-1}[\mathbf{x}(0) + B\mathbf{U}(s)]$$

For the transfer functions, we are not really interested in the initial conditions of $\mathbf{x}$; we assume that they are zero when the input is applied, or that a sinusoidal output has settled to a steady state. The equation can be reduced to

$$\mathbf{Y}(s) = C(sI - A)^{-1}B\mathbf{U}(s) \tag{8-3}$$

The matrix

$$C(sI - A)^{-1}B$$

is a matrix of transfer functions having as many rows as there are outputs, and as many columns as there are inputs.

Remember that in inverting $sI - A$, we will have had to evaluate its determinant, the characteristic polynomial of A, and this will appear in the denominator. In other words, the roots of the characteristic polynomial provide the poles of the system as it stands, before the application of extra feedback. Do not expect, however, that each of the transfer function terms will have all these poles; in many cases they will cancel out with factors of the numerator polynomials. However, we can assert that the poles of the transfer functions can only come from the roots of the characteristic equation.

Exercise 8-3-1 *Use this method to derive a transfer function for the example defined by Eqs (8-1).*

Exercise 8-3-2 *Derive the transfer function matrix for*

$$\begin{bmatrix} \dot{x}_1 \\ \dot{x}_2 \end{bmatrix} = \begin{bmatrix} -3 & 0 \\ 0 & -2 \end{bmatrix} \begin{bmatrix} x_1 \\ x_2 \end{bmatrix} + \begin{bmatrix} 1 & 0 \\ 0 & 1 \end{bmatrix} \begin{bmatrix} u_1 \\ u_2 \end{bmatrix}$$

$$y = \begin{bmatrix} 1 & 1 \end{bmatrix} \begin{bmatrix} x_1 \\ x_2 \end{bmatrix}$$

8-4 TRANSFER FUNCTIONS AND TIME RESPONSES

In the mathematical prelude to this chapter, the impulse response was mentioned. Given a relationship between an input and an output defined by the transfer function $H(s)$, the application of a unit impulse at the input will provoke an output $h(t)$, where $H(s)$ is the Laplace transform of $h(t)$.

Suppose that $H(s) = 1/(s + a)$. A trip to the tables of Laplace transforms will tell us that $h(t) = \exp(-at)$, in other words the immediate response has value unity, decaying to zero with time constant $1/a$. Is there some way of gaining insight into the response function without actually looking up or working out the inverse transform?

There are two powerful properties of the Laplace transform which will help us, the *initial value theorem* and the *final value theorem*.

The initial value theorem tells us that

$$\lim_{t \to 0} f(t) = \lim_{s \to \infty} sF(s)$$

while the final value theorem states that

$$\lim_{t \to \infty} f(t) = \lim_{s \to 0} sF(s)$$

Rather than construct a formal proof, let us look for a plausible argument. If $F(s)$ is a ratio of polynomials, we can rearrange it as a ratio of polynomials in $1/s$. We can then perform a 'long division' to obtain a power series in $1/s$.

Thus $1/(s + 1)$ would become $(1/s)/(1 + 1/s)$, which we would divide out to obtain

$$\frac{1}{s} - \left(\frac{1}{s}\right)^2 + \left(\frac{1}{s}\right)^3 \cdots$$

The inverse transform of this series will be a power series in t, since $1/s^{n+1}$ is the Laplace transform of $t^n/n!$. If we are only interested in the value at $t = 0$, we can ignore all terms beyond the $1/s$ term. Now if there is a constant at the start of the series, we have an infinite impulse on our hands. If it starts with a $1/s$ term, we have a respectable step, while if the first term is of higher power then our initial value is zero.

To extract the $1/s$ term's value, all we need to do is multiply the series by s and let s tend to infinity—if there is a constant term, we will correctly get an infinite result. But why go to the trouble of dividing out? We need only multiply $F(s)$ by s, let s tend to infinity and there is the initial value of $f(t)$.

The next argument to demonstrate the final value theorem is even more woolly!

If we have a relationship $Y(s) = H(s)U(s)$, and if $H(s)$ is the ratio of two polynomials in $sZ(s)/P(s)$, then we have

$$P(s)Y(s) = Z(s)U(s)$$

This relationship can be construed as meaning that $y(t)$ and $u(t)$ are linked by a differential equation

$$P(D)y(t) = Z(D)u(t)$$

where D represents the operator d/dt. If we assume that everything is stable and that the $u(t)$ that is applied is the unit step, then eventually all the derivatives of y will become zero, while the derivatives of u will only have served to provoke initial transients. Thus for the limiting value of y as t tends to infinity we can let D tend to zero in the polynomials. This is just the same as letting s tend to zero, so we can reassemble $H(s)$ to state that the following. If a unit step is applied to a transfer function $H(s)$, then as time tends to infinity the output will tend to

$$\lim_{s \to 0} H(s)$$

The unit impulse is the time derivative of the unit step, so (begging all sorts of questions of convergence) the output for impulse input will be the derivative of the output for a step input. We can serve up a derivative by multiplying $H(s)$ by another s, obtaining

$$\lim_{t \to \infty} h(t) = \lim_{s \to 0} sH(s)$$

Well, I warned you!

In fact it is often more helpful to look at the step response of a filter than

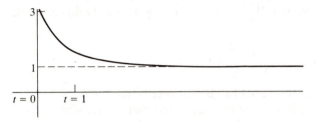

Figure 8-1 Step response of a phase advance filter $(1 + 3s)/(1 + s)$.

at the impulse response. Consider the phase advance element with the transfer function

$$H(s) = \frac{1 + 3s}{1 + s}$$

For the step response initial and final values, we must look at the result of applying $1/s$ to the transfer function $H(s)$. Instead of taking the limits of $sH(s)$, the input $1/s$ will cancel with the 'spare' s to leave us with the limits of $H(s)$ as s tends to infinity and as s tends to zero respectively.

For the initial value of the step response, we let s tend to infinity in $H(s)$, which in this case gives us the value 3. For the final value of the step response, we let s tend to zero in $H(s)$, here giving the value 1. From the denominator we expect the function $\exp(-t)$ to be involved, so we can intelligently sketch the step response as in Fig. 8-1.

We can add step responses together to visualize the results of various transfer functions. The phase advance we have just considered, for instance, can be split into two terms:

$$\frac{1 + 3s}{1 + s} = 1 + \frac{2s}{1 + s}$$

We can generate a filter to produce this response by mixing a 'straight-

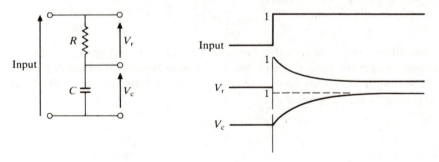

Figure 8-2 Voltage waveforms making up step response of an R–C circuit.

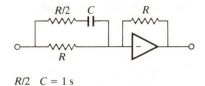

Figure 8-3 Operational-amplifier circuit with gain $-(1 + 3s)/(1 + s)$.

through' term having a gain of unity with another signal which gives the second term above. Now this second term's step response has the initial value 2 and final value zero. If we apply a step of voltage across the R–C circuit of Fig. 8-2, the voltage across the resistor will be filtered by $s/(1 + s)$, giving the transient shown in the diagram. To construct the phase advance we need to add twice this transient to the basic step, and Fig. 8-3 shows a suitably simple circuit to generate the desired result.

In passing, note that in the simple circuit of Fig. 8-2 the resistor voltage is filtered by $s/(1 + s)$ while the capacitor voltage is filtered by $1/(1 + s)$, a 'lag' which smoothes the response by attenuating higher frequencies. When these responses are added together, we get back to the original input. In other words, we can construct the transient response by subtracting a lagged signal from the original signal.

8-5 FILTERS IN SOFTWARE

Earlier on, we saw that a dynamic system could be simulated by means of its state equations. When a signal is applied to the system, an output is obtained which is related to the input by means of a transfer function. When we apply the same input to our simulation, if all is well we will obtain the same output function as in the real system.

A filter is merely the means of applying a transfer function to an input signal. It might be constructed from resistors, capacitors and amplifiers or it might contain pneumatic reservoirs. The implementation is of secondary concern; it is the effect on the input signal that matters.

If we simulate such a filter, we have not just got a simulation, we have constructed an alternative filter which we could use in place of the original (assuming, of course, that the input and output signals have the correct qualities).

Let us look yet again at the task of constructing the phase advance element $(1 + 3s)/(1 + s)$, this time using a few lines of software in a computer program.

The first task is to set up some state equations. The denominator suggests that all we need is

$$\dot{x}_1 = -x_1 + u$$

to give us the 'lag' $X_1(s) = U(s)/(s + 1)$. When we try to set up an output in the form

$$y = Cx$$

however, we immediately find problems. The state vector only has one element, which gives a lagged version of the input. No amount of multiplication by constants will sharpen it into a phase advance. Remember, however, that a transient can be obtained by subtracting a lag from the original input. If we settle for

$$y = 3u - 2x_1$$

we see that $Y(s) = [3 - 2/(1 + s)]U(s)$, which reduces to the desired function. To achieve a transfer function with as many zeros as it has poles, it is clear that we must broaden our output definition to

$$y = Cx + Du$$

Why was the output not defined in this way in the first place? When we come to apply feedback around a system which has 'instant response', we must solve simultaneous algebraic equations as well as differential equations to find the output. To produce generalized theorems, it is safer to stick to the simpler form of output and to treat this form as an exception whenever it cannot be avoided.

So now we have two equations, which can be represented by the computer assignment statements

```
DX1 = U − X1
```

and

```
Y = 3*U − 2*X1
```

We can introduce the time element by the simplest of integrations:

```
X1 = X1 + DX1*DT
```

and all the ingredients are present for the software model. Suppose we settle for a time increment of 0.01 seconds; then these three lines can be simplified even further to give

```
X1 = X1 + .01*(U − X1)
Y = 3*U − 2*X
```

If we can be sure that these lines will be executed one hundred times per second, then we have a real-time filter as specified. If the filter is to be part of an off-line model, we only have to ensure that the model time advances by 0.01 seconds between executions.

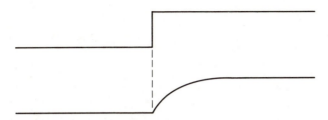

Figure 8-4 Effect of a low-pass filter on a step waveform.

Suppose that we have an array of sampled results, stored in V(1) to V(200). We can smooth these rather drastically with

```
N = 50
X = V(1)
FOR I = 2 TO 200
X = X + (V(I) − X)/N
V(I) = X
NEXT I
```

The effect is as though the signal had been subjected to a low-pass filter with the time constant equal to 50 times the sample interval. Decrease the value of N and the filter time constant will reduce. If the array is set up to represent a signal with a step change and the values are plotted before and after filtering, then the result will appear as in Fig. 8-4. The values have been 'smeared' to the right.

In the case of the stored array of samples, we could smear the results to the left instead by performing the computation in the reverse order. If we apply both filters, however, we can balance the smearing out completely:

```
N = 50
X = V(1)
FOR I = 2 TO 200
X = X + (V(I) − X)/N
V(I) = X
NEXT I
FOR I = 199 TO 1 STEP −1
X = X + (V(I) − X)/N
V(I) = X
NEXT I
```

The result is shown in Fig. 8-5. It is equivalent to processing by the second-order filter $1/[1 − (\tau s)^2]$. Writing $j\omega$ for s, we find the frequency response to be $1/(1 + \tau^2\omega^2)$, which is always real; this filter has zero phase shift at all frequencies.

To apply such a filter to a real-time signal is of course out of the question.

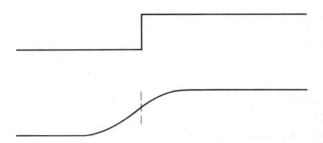

Figure 8-5 The filter $1/(1 + \tau^2 s^2)$ has a symmetrical effect in time.

One of the poles is hopelessly unstable and it is only because we have applied its contribution in 'negative time' that we have been able to get away with the trick. Look again at the way the filtered signal starts to change before the step arrives, and you will see that the filter is *non-causal*—the output depends not just on the past values of the input signal but also on its future.

8-6 STATE EQUATIONS IN THE COMPANION FORM

If the required filter is causal, after all, and if we have a transfer function to specify it, how can we arrive at a suitable set of state equations? Suppose that the transfer function is the ratio of two polynomials, $Z(s)/P(s)$, where as before $Z(s)$ defines the zeros while $P(s)$ defines the poles. Take a practical example,

$$Z(s) = 4s^2 + s + 1$$

and

$$P(s) = s^3 + 3s^2 + 4s + 2 \tag{8-4}$$

We can build a set of state equations in 'companion form' as follows. If we make the second state variable the derivative of the first, and the third the derivative of the second and so on, then we have

$$\dot{x}_1 = x_2$$

$$\dot{x}_2 = x_3$$

where the state variables represent successively higher derivatives of the first. Now it is clear that

$$\dot{x}_3 + 3x_3 + 4x_2 + 2x_1 \tag{8-5}$$

will be equivalent to

$$\dddot{x}_1 + 3\ddot{x}_1 + 4\dot{x}_1 + 2x_1$$

If we equate this to the input u and take the Laplace transform, we have

$$(s^3 + 3s^2 + 4s + 2)X_1(s) = U(s)$$

We have a system with the correct set of poles. We must rearrange expression (8-5) into a format more suitable for a state equation, and we get

$$\dot{x}_3 = -2x_1 - 4x_2 - 3x_3 + u$$

In matrix form, the state equations now become

$$\dot{\mathbf{x}} = \begin{bmatrix} 0 & 1 & 0 \\ 0 & 0 & 1 \\ -2 & -4 & -3 \end{bmatrix} \mathbf{x} + \begin{bmatrix} 0 \\ 0 \\ 1 \end{bmatrix} u$$

That settles the denominator. How do we arrange the zeros, though? From the way x_2 was defined, we have $X_2(s) = sX_1(s)$. Similarly, x_3 is the second derivative, so $X_3(s) = s^2 X_1(s)$. We can immediately generate the numerator polynomial by setting

$$y = 4x_3 + x_2 + x_1$$

$$= \begin{bmatrix} 1 & 1 & 4 \end{bmatrix} \mathbf{x}$$

Once again, we can only get away with this form, $y = C\mathbf{x}$, if there are more poles than zeros. If they are equal in number, first perform one stage of 'long division' of the numerator polynomial by the denominator to isolate the term proportional to the input, and the remainder will fit into the pattern. If there are more zeros than poles, give up.

Whether it is a simulation or a filter, the system is represented by a few lines of software. If we were to turn over a stone, we would find under it a lot of unanswered questions about the stability of the simulation, about the quality of the approximation and about the choice of step length. Those will all be answered in a later chapter on discrete-time systems. For now let us turn our attention to the computational techniques of convolution.

Exercise 8-6-1 *We wish to synthesize the filter $s^2/(s^2 + 2s + 1)$ in software. Set up the state equations and write a brief segment of program.*

8-7 DELAYS AND THE UNIT IMPULSE

We have already looked into the function of time which has a Laplace transform that is just 1, the *delta function* $\delta(t)$. The unit step has the Laplace transform $1/s$, and so we can think of the delta function as its derivative. Before we go on, we must derive an important property of the Laplace transform, the *shift theorem*.

If we have a function of time, $x(t)$, which is zero for all $t < 0$, and if we pass this signal through a time delay τ, then the output will be $x(t - \tau)$. The Laplace transform of this output will be

$$\int_{t=0}^{\infty} x(t - \tau) \, e^{-st} \, dt$$

If we write T for $t - \tau$, then dt will equal dT, T will have the value $-\tau$ at $t = 0$ and the integral becomes

$$\int_{T=-\tau}^{\infty} x(T) \, e^{-s(T+\tau)} \, dT = e^{-s\tau} \int_{T=-\tau}^{\infty} x(T) \, e^{-sT} \, dT$$

However $x(T)$ is zero for negative T, so we may change the integration limits to be between $T = 0$ and infinity. Now the integral reveals itself to be just the Laplace transform of $x(t)$, with the extra multiplier $e^{-s\tau}$. We have

$$\mathscr{L}\{x(t - \tau)\} = e^{-s\tau} \, X(s)$$

Matters could have become awkward if instead of delaying the signal, we had wanted to advance it in time; we could not so easily have changed the limits of integration. We can relax the condition that the signal must be zero for all negative time by redefining the Laplace transform to be the *bilateral transform*, integrated from $t = -\infty$ to $+\infty$ rather than from $t = 0$. Clearly signals have to quieten down for negative t, otherwise their contribution to the integral would be enormous when multiplied by the exponential. Now we can allow time advances within reason, as well as delays, without having to resort to special treatment.

Needless to say, the time function with unity transform is unchanged by this redefinition; it is still the impulse $\delta(t)$ having infinite value at $t = 0$.

We can immediately start to put the shift theorem to use; it tells us that the transform of $\delta(t - \tau)$, the unit impulse shifted to occur at $t = \tau$, is $e^{-s\tau}$. We could, of course, have worked this out from first principles. Sticking with the bilateral transform, we have

$$\mathscr{L}\{\delta(t - \tau)\} = \int_{t=-\infty}^{\infty} \delta(t - \tau) \, e^{-st} \, dt$$

Now $\delta(t - \tau)$ is zero everywhere except at $t = \tau$. The integral of its impulse at $t = \tau$ is unity, but here the impulse is multiplied at that value of t by $e^{-s\tau}$. We can deduce that the value of the transform integral is

$$e^{-s\tau}$$

In effect, the action of multiplying the function e^{-st} by the unit impulse at $t = \tau$ and integrating over all time has just taken a sample of the exponential at $t = \tau$. This will be true of any function of time:

$$\int_{t=-\infty}^{\infty} x(t) \, \delta(t - \tau) \, dt = x(\tau) \tag{8-6}$$

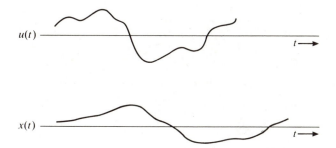

Figure 8-6 $u(t)$ and $x(t)$ are regarded as time histories, not just as instantaneous values.

Let us briefly indulge in a little philosophy about the 'meaning' of functions. We can think of $x(t)$ as a simple number, the result of substituting some value of t into a formula for computing x. We can instead expand our vision of the function to look at the graph of $x(t)$ plotted against time, as in a step response. In control theory we have to take this broader view, regarding inputs and outputs as time histories and not just as simple values (see Fig. 8-6).

We can look upon Eq. (8-6) as a sampling process, allowing us to pick out of the time history one single value of the function. However if we exchange the symbols t and τ in the equation suddenly the perspective changes. The substitution has no absolute mathematical effect, but if we come to regard $x(t)$ as a time history and $x(\tau)$ as a single value, then

$$\int_{\tau = -\infty}^{\infty} x(\tau)\, \delta(\tau - t)\, d\tau \tag{8-7}$$

which tells us something quite different from Eq. (8-6). It suggests that we can regard the time history $x(t)$ as being made up of an infinite train of infinitesimal impulses. At each time τ, the impulse has magnitude $x(\tau)\, d\tau$.

This result may not look important, but it opens up a whole new way of looking at the response of a system to an applied input.

8-8 THE CONVOLUTION INTEGRAL

Let us first define the situation. We have a system described by a transfer function $G(s)$, with input function $u(t)$ and output $y(t)$, as in Fig. 8-7. If we

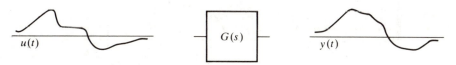

Figure 8-7 $G(s)$ gives output $y(t)$ for input $u(t)$.

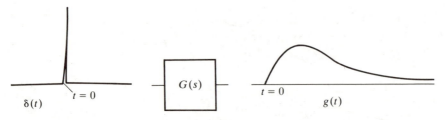

Figure 8-8 For unit impulse input, $G(s)$ gives $g(t)$.

apply a unit impulse to the system at $t = 0$, the output will be $g(t)$, where the Laplace transform of $g(t)$ is $G(s)$, as in Fig. 8-8. How do we go about deducing the output function for any general $u(t)$?

Perhaps the most fundamental property of a linear system is the *principle of superposition*; if we know the output response to one input function and to another input function, then if we add the two input functions together and apply them, the output will be the sum of the two corresponding output responses.

In mathematical terms, if $u_1(t)$ produces the response $y_1(t)$ and $u_2(t)$ produces the response $y_2(t)$, then an input of $u_1(t) + u_2(t)$ will give an output of $y_1(t) + y_2(t)$.

An input $\delta(t)$ to $G(s)$ provokes the output $g(t)$. An impulse $u(\tau)\,\delta(t - \tau)$ applied at time $t = \tau$ gives the delayed response $u(\tau)g(t - \tau)$. If we apply several impulses in succession, the output will be the sum of the individual

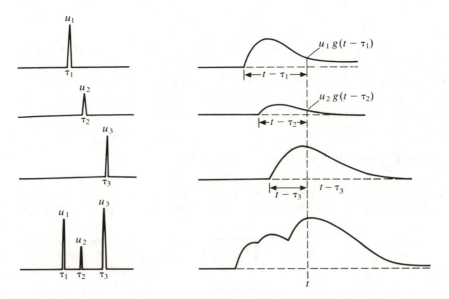

Figure 8-9 The response is the sum of the responses of the individual impulses.

responses, as shown in Fig. 8-9. If we inspect the output at some later time t, we see that the effect of the first impulse has already 'grown old', decaying to the value $u(\tau_1)g(t - \tau_1)$. The latest impulse has an effect that is still fresh; as the term in the u bracket increases, the term in the g bracket reduces.

We should now put matters onto a formal basis. The response to a single unit impulse at time τ is $g(t - \tau)$. Provided the necessary conditions are satisfied for convergence of the integral, the response to an infinite train of impulses $u(\tau)\,\delta(t - \tau)\,d\tau$ will by superposition be seen to be

$$y(t) = \int_{\tau = -\infty}^{\infty} u(\tau)g(t - \tau)\,d\tau$$

This is the convolution integral. The limits are infinite in the general case, but when we know more about $u(t)$ and $g(t)$ we can make them more manageable. If $u(\tau)$ is zero for negative τ, then the lower limit can be made zero. If $g(t)$ is also zero for negative t (the system is causal) then the upper limit of τ can be made equal to t.

8-9 FINITE IMPULSE RESPONSE FILTERS

We see that instead of simulating a system to generate a filter response, we could set up an impulse response time function and produce the same result by convolution. With infinite integrals lurking around the corner, this might not seem such a wise way to proceed!

In looking at digital simulation, we have already cut corners by taking a finite step length and accepting the resulting approximation. A digital filter must similarly accept limitations in its performance in exchange for simplification. Instead of an infinite train of impulses, $u(t)$ is now viewed as a train of samples at finite intervals. The infinitesimal $u(\tau)\,d\tau$ has beome $u(n\tau)\tau$. Instead of impulses, we have numbers to input into a computational process.

Similarly, $g(t)$ is broken down into a train of sample values, using the same sampling interval. Now the infinitesimal operations of integration are coarsened into the summation

$$y(n\tau) = \sum_{r = -\infty}^{r = \infty} \tau u(r\tau)g((n - r)\tau)$$

The infinite limits still do not look very attractive. For a causal system, however, we need go no higher than n, while if the first signal was applied at $r = 0$ then this can be the lower limit.

Summing from $r = 0$ to n is a definite improvement, but means that we have to sum an increasing number of terms as time advances. Can we do any better?

Most filters will have a response that eventually decays after the application of an impulse. The one-second lag $1/(s + 1)$ has an initial

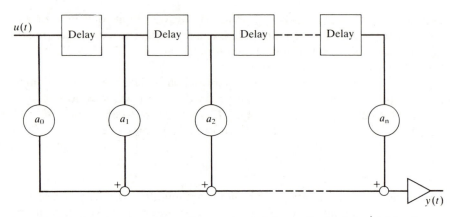

Figure 8-10 A finite impulse response filter can be made from a 'bucket-brigade' delay line.

response of unity, gives an output of around 0.37 after one second, but after ten seconds the output has decayed to less than 0.00005. There is a point where $g(t)$ can safely be ignored, where indeed it is less than the resolution of the computation process. Instead of regarding the impulse response as a function of infinite duration, we can cut it short to become a *finite impulse response*. This is the basis of the FIR filter.

Now we may say that $g(r\tau)$ is zero for all $r < 0$ and for all $r > N$. We have

$$y(n\tau) = \tau \sum_{r=n-N}^{r=n} u(r\tau)g((n-r)\tau)$$

We can tidy this up by writing $n - r$ in place of r, to get

$$y(n\tau) = \tau \sum_{r=0}^{r=N} u((n-r)\tau)g(r\tau)$$

The output now depends on the input u at the time in question, and on its past N values. These values are now multiplied by appropriate fixed coefficients and summed to form the output, and are moved along one place to admit the next input sample value. The method lends itself ideally to a hardware application with a 'bucket-brigade' delay line, as shown in Fig. 8-10.

The following software suggestion can be made much more efficient in time and storage; it concentrates on showing the method. Assume that the impulse response has already been set up in the array $G(I)$, where I ranges from 0 to N. We provide another array $U(I)$ of the same length to hold past values.

```
REM MOVE SAMPLES TO MAKE ROOM
FOR I = N TO 1 STEP −1
U(I) = U(I − 1)
NEXT1
```

```
REM TAKE IN NEW SAMPLE
U(0) = NEWSAMPLE

REM NOW COMPUTE OUTPUT
Y = 0
FOR I = 0 TO N
Y = Y + U(I)*G(I)
NEXT I
```

Y now holds the current output value.

This still seems more trouble than the simulation method; what are the advantages? First, there is no question of the process becoming unstable. Extremely sharp filters can be made for frequency selection or rejection which would have poles very close to the stability limit; since the impulse response is defined exactly, stability is assured.

Next, the rules of causality can be bent a little. Of course, the output cannot precede the input, but by adding a delay to the main output signal the impulse response can have a 'leading tail'. Take the non-causal smoothing filter discussed earlier, for example. This has a bell-shaped impulse response, symmetrical about $t = 0$, as shown in Fig. 8-11. By delaying this function, all the important terms can be contained in a positive range of t. There are many applications, such as sound and picture filtering, where the added delay is no embarrassment.

$g(t)$

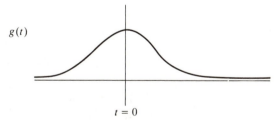

$t = 0$

Non-causal response, impossible in real time

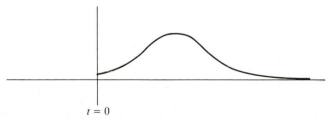

$t = 0$

A time shift makes a causal approximation

Figure 8-11 By delaying a non-causal response, it can be made causal.

8-10 CONCLUSION

We have seen that the state description of a system and its representation as an array of transfer functions are intimately bound together. We have requirements that appear to conflict. On the one hand, we seek a formalism which will allow as much of the work as possible to be undertaken by computer. On the other, we wish to retain an insight into the nature of the system and its problems, so that we can use intelligence in devising a solution. Do we learn more from the time domain, trading in matrix equations and step and impulse responses, or does the transfer function tell us more, with its implied frequency response and root locus?

In the next chapter, we will start to tear the state equations apart to see what the system is made of. Maybe we can get the best of both worlds.

EIGENVECTORS AND FUNCTIONS OF A MATRIX

P9-1 INTRODUCTION

Before taking a can-opener to the matrix state equations, we must look at some of the properties of a matrix that are unchanged by a transformation of variables. We also need to consider such unlikely functions of a matrix as the exponential. First of all, however, we must see what is meant by the *rank* of a matrix.

P9-2 THE RANK OF A MATRIX

There are two ways of looking at the operation of multiplying a matrix by a vector. Recall that the product of a 3 by 3 example can be expanded as follows:

$$
\begin{bmatrix} a & b & c \\ d & e & f \\ g & h & i \end{bmatrix} \begin{bmatrix} p \\ q \\ r \end{bmatrix} = \begin{bmatrix} ap + bq + cr \\ dp + eq + fr \\ gp + hq + ir \end{bmatrix}
$$

Each element of the result can be regarded as the scalar product of a row of A with the multiplying vector

$$
\begin{bmatrix} (a, b, c)(p, q, r)' \\ (d, e, f)(p, q, r)' \\ (g, h, i)(p, q, r)' \end{bmatrix}
$$

The vector $(p, q, r)'$ is resolved in the directions of the three row vectors making up matrix A. (Remember, by the way, that the prime denotes the transpose of a vector or matrix. It is easier to write row vectors and mark them as transposed into column vectors than to take up acres of paper and lose the thread of the argument.)

A slight change of perspective reveals the product as a mixture of the columns of A:

$$\begin{bmatrix} a \\ d \\ g \end{bmatrix} p + \begin{bmatrix} b \\ e \\ h \end{bmatrix} q + \begin{bmatrix} c \\ f \\ i \end{bmatrix} r$$

We will start with this second form to explore the concept of rank. Let us name the three column vectors which make up A as **a**, **b** and **c**. Suppose that **b** and **c** are multiples of **a** so that all three vectors point in the same direction. Then any mixture of these vectors must also lie in the same direction; however varied we make **p**, the answer has only one degree of freedom. In this case, we say that the rank of A is just 1. We see also that we can find many combinations of the values of the elements of **p** that will cause the contributions to cancel out, to give a result of $(0, 0, 0)'$.

We may define rank as the number of columns of A that are *linearly independent*. If all three are again multiples of each other, then any two can be mixed in a combination which will cancel to $(0, 0, 0)'$. The *column rank* is the number of linearly independent columns of A. It can be shown that the number of linearly independent columns is also equal to the number of linearly independent rows, the *row rank*. If this rank is just 1, then any vector 'orthogonal' to this single row vector direction will have all three scalar products zero; the result of multiplying it by A will be an all-zero vector.

Now let us consider a rank of two. The deficiency might be obvious, with one row or column consisting only of zero coefficients, but it is likely to be more subtle. Any two vectors of A might look completely healthy, but when all three are considered together they are perhaps seen to lie in a plane. The third vector can be made up from a combination of the other two.

Now in our second view of the matrix product, it is clear that, try as we might, we cannot find a vector **p** that will give a resulting mix of column vectors that has a component orthogonal to the plane in which the column vectors lie. The result is just a flat 'shadow' of any **p** cast onto that plane. From the scalar product point of view, if **p** is orthogonal to the plane of the row vectors of A then all three scalar products are again zero. This orthogonal direction is the *null space* of A; the product of A with a vector in its null space is zero.

For A to have rank 3, all its rows must be linearly independent, as must be all its columns. The three row vectors can become a coordinate system against which an unknown vector can be resolved, such that no two different vectors will give the same set of scalar products. The three column vectors can equally be regarded as a coordinate system, such that any particular vector can be made up from a unique mixture of the three columns.

We can go a step further if the rank of A is only two, by choosing the null-space vector as one vector of a coordinate system together with two other linearly independent vectors of A. If the rank of A is just one, then there are two dimensions to the null space.

How do we test the rank of the matrix? It is easy to tell if its rank is equal to its number of dimensions, three in this case. If the resolving or mixing process loses no information, then the process is reversible; matrix A has an inverse. We saw, too, that a necessary condition for the existence of an inverse was that the determinant must be non-zero.

A non-zero determinant is also a sufficient condition. If we can find a null space vector, this implies that a combination of the columns can give $(0, 0, 0)'$. We can mix combinations of columns without changing the value of the determinant. If we can fabricate an all-zero column by a legitimate mixture then the determinant of A must be zero. If existence of a null space implies a zero determinant, then a non-zero determinant proves that there is no null space; the matrix has full rank.

Although a matrix with less than full rank cannot be inverted, a 'sort of' inverse can be constructed by the elimination method of the prelude to the last chapter. A complete set of diagonal ones cannot be achieved—just as many as the rank of the matrix, with zeroes completing the matrix.

P9-3 MATRIX TRANSFORMATIONS

In Chapter 3 we saw some interesting results of changing the coordinate system in which the state $\mathbf{x}$ was measured. When the new vector $\mathbf{w}$ was used, where

$$\mathbf{w} = T\mathbf{x}$$

the state equation

$$\dot{\mathbf{x}} = A\mathbf{x} + B\mathbf{u}$$

became

$$\dot{\mathbf{w}} = TAT^{-1}\mathbf{w} + TB\mathbf{u}$$

The new equations still represent the same system, so the 'essential' properties must be unchanged by any valid transformation. What are they?

Let us take as an example the matrix

$$A = \begin{bmatrix} 2 & 2 \\ 3 & 1 \end{bmatrix}$$

If we postmultiply this by the column vector $(1, 0)'$ we get another column vector $(2, 3)'$. The direction of the new vector is different and its size is changed. Premultiplying by A takes $(0, 1)'$ to $(2, 1)'$, takes $(1, -1)'$ to $(0, 2)'$ and takes $(1, 1)'$ to $(4, 4)'$. Wait a minute, though. Is there something special about this last example? The vector $(4, 4)'$ is in exactly the same direction as the vector $(1, 1)'$, and is multiplied in value by 4. Whatever set of coordinates we choose, the property that there is a vector on which A just has the effect of multiplication by 4 will be unchanged.

Is this the only vector on which A has this sort of effect? We can easily find out. We are looking for a vector $\mathbf{x}$ such that $A\mathbf{x} = \lambda\mathbf{x}$, where λ is a mere constant. Now

$$A\mathbf{x} = \lambda\mathbf{x} = \lambda I\mathbf{x}$$

so

$$(\lambda I - A)\mathbf{x} = \begin{bmatrix} 0 \\ 0 \end{bmatrix} \tag{P9-1}$$

We have a respectable vector $\mathbf{x}$ being multiplied by a matrix to give a null vector. Obviously the matrix is deficient in rank, and its determinant is zero:

$$\det(\lambda I - A) = 0$$

Write s in place of λ and this should bring back memories. How that characteristic polynomial gets around!

In the example above, the determinant is

$$\det \begin{vmatrix} \lambda - 2 & -2 \\ -3 & \lambda - 1 \end{vmatrix}$$

giving

$$\lambda^2 - 3\lambda - 4 = 0$$

which factorizes into

$$(\lambda + 1)(\lambda - 4) = 0$$

The root $\lambda = 4$ comes as no surprise, and we know that it corresponds to the *eigenvector* $(1, 1)'$. Let us find the vector that corresponds to the other value, -1. Take this value for λ and substitute it into Eq. (P9-1):

$$\begin{bmatrix} -3 & -2 \\ -3 & -2 \end{bmatrix} \begin{bmatrix} x \\ y \end{bmatrix} = \begin{bmatrix} 0 \\ 0 \end{bmatrix}$$

It is no surprise that the two simultaneous equations have degenerated into one (if they had not, we could not solve for a single value of x/y). The vector $(2, -3)'$ is obviously a solution, as is any multiple of it. Multiply by A and we are reassured to see that the result is $(-2, 3)$.

In general, if A is n by n we will have n roots to the characteristic equation, and we should be able to find n eigenvectors to go with them. If the roots, or *eigenvalues*, are all different, it can be shown that the eigenvectors are all linearly independent. We can stack them up into a respectable transformation matrix, T^{-1}, which when premultiplied by A will give a stack of columns each the original eigenvector now multiplied by its eigenvalue. Premultiply this by the inverse, T, and we arrive at the elegant result of a diagonal matrix, each element of which is one of the eigenvalues.

To sum up, solve

$$\det(\lambda I - A) = 0$$

to find the eigenvalues. Substitute the n eigenvalues in turn into

$$(\lambda I - A)\mathbf{x} = \mathbf{0}$$

to find the n eigenvectors in the form of column vectors. Stack these column vectors together, in order of decreasing eigenvalue for neatness, to make a matrix T^{-1}. Find its inverse, T. We now see that

$$TAT^{-1} = \begin{bmatrix} \lambda_1 & 0 & 0 & \cdots \\ 0 & \lambda_2 & 0 & \cdots \\ 0 & 0 & \lambda_3 & \cdots \\ \multicolumn{4}{c}{\cdots\cdots\cdots\cdots\cdots} \end{bmatrix}$$

Exercise 9-3-1 *Perform the above operations on the matrix*

$$\begin{bmatrix} 0 & 1 \\ -6 & -5 \end{bmatrix}$$

Then look back at Sect. 3-4.

When we have repeated roots, there may be problems. They will be dealt with in the main chapter. For now, let us look at functions of a matrix.

P9-4 EXPONENTIAL FUNCTION OF A MATRIX

Some functions of a matrix are pretty obvious. The square of A is found just by multiplying A by itself. The cube requires another multiplication by A, and the same goes for any integer power within reason. We have met the inverse, so negative powers follow simply. How should we approach anything so strange as the exponential of a matrix?

To be more precise, the function we would like to define is the exponential of At, where t is a scalar that represents time. The property we are looking for is

$$\frac{d}{dt}(e^{At}) = A\, e^{At}$$

to help in solving matrix differential equations. Now if instead we consider the more humble $\exp(at)$, where a is a mere scalar, we can expand it as a power series in t:

$$e^{at} = 1 + at + \frac{(at)^2}{2!} + \frac{(at)^3}{3!} + \cdots$$

So powerful is the factorial in the denominator that it can make the summation of the series converge for any finite value of at.

We can demonstrate the property of the derivative of this function by differentiating term by term, to get

$$\frac{d}{dt}(e^{at}) = a + 2a\frac{(at)}{2!} + 3a\frac{(at)^2}{3!} + \cdots$$

$$= a + a(at) + a\frac{(at)^2}{2!} + \cdots$$

$$= a\left[1 + at + \frac{(at)^2}{2!} + \cdots\right]$$

$$= a\, e^{at}$$

So why not try the same trick, but with the matrix A in place of the scalar a? We have to use the unit matrix I in place of the value 1, of course, but the argument is the same. To make the summation, we multiply each element of each contributing matrix by the appropriate power of t, divide by the corresponding factorial constant and add it to the running total for that element position.

The summation must converge for each element, as seen from the following argument. If the sum of the magnitudes of all the coefficients of A is M, then no element of A^n can exceed a value of M^n. If the series for e^{Mt} converges, which is the case for every finite Mt, then so must the series for each element of the exponential of At. If

$$e^{At} = I + At + \frac{(At)^2}{2!} + \frac{(At)^3}{3!} + \cdots$$

then

$$\frac{\mathrm{d}}{\mathrm{d}t}(\mathrm{e}^{At}) = A + 2A\frac{At}{2!} + 3A\frac{(At)^2}{3!} + \cdots$$

$$= A + A(At) + A\frac{(At)^2}{2!} + \cdots$$

$$= A\left[I + At + \frac{(At)^2}{2!} + \cdots\right]$$

$$= A\,\mathrm{e}^{At}$$

Since we must be fussy about the order of multiplication of matrices, we should also note at this time that the factor A could be drawn to the right of each term of this series, to show that

$$\frac{\mathrm{d}}{\mathrm{d}t}(\mathrm{e}^{At}) = \mathrm{e}^{At}\,A$$

We can use the results of the last section to tidy up all such functions rather neatly (if the eigenvalues are all different). We saw that a transformation T could be found, so that

$$TAT^{-1} = \Lambda$$

Here Λ is a diagonal matrix, having all elements zero except the diagonal elements which take the values of the eigenvalues.

If we were looking for functions of Λ instead of A, life would be much easier. With little effort (exercise for the reader!) it can be shown that $\Lambda\Lambda$ is another diagonal matrix, the diagonal elements this time being the squares of the eigenvalues. The *nth* power of Λ has elements that are the *n*th powers of the eigenvalues, and it takes little imagination to see that the exponential of (Λt) will be a diagonal matrix with elements of the form $\exp(\lambda t)$.

Now if

$$TAT^{-1} = \Lambda$$

we have the corresponding property that A can be expressed as

$$A = T^{-1}\Lambda T$$

Now

$$A^2 = AA$$

$$= T^{-1}\Lambda TT^{-1}\Lambda T$$

$$= T^{-1}\Lambda I\Lambda T$$

$$= T^{-1}\Lambda^2 T$$

A similar result holds for each power of A. We can take out factors of T^{-1} and T before and after the power series, and our series in A becomes a corresponding power series in Λ. We are led at once to the conclusion that

$$e^{At} = T^{-1} e^{\Lambda t} T$$

$$= T^{-1} \begin{bmatrix} e^{\lambda_1 t} & 0 & 0 & \cdots \\ 0 & e^{\lambda_2 t} & 0 & \cdots \\ 0 & 0 & e^{\lambda_3 t} & \cdots \\ & & \cdots & \end{bmatrix} T$$

Let us see how these results can be applied to the state equations.

MORE ABOUT TIME AND STATE EQUATIONS

9-1 INTRODUCTION

We have now seen the same system described by arrays of transfer functions, differential equations and first-order matrix state equations. We have seen that, however grandiose the system may be, the secret of its behaviour is unlocked by finding the roots of a single polynomial characteristic equation. The complicated time solution then usually crumbles into an assortment of exponential functions of time.

In this chapter we are going to tackle the matrix state equations using algebra to open up and simplify the structure. We will see that a transformation of variables will let us unravel the system into an assortment of simple subsystems, whose only interaction occurs at the input or the output.

In switching to and fro between state representation and transfer functions, a system such as a single lag can be depicted in several ways. At one moment it is drawn as a box labelled $1/(s + a)$; the next it is shown as an analogue integrator with feedback a around it. The analogue integrator may then vanish to be replaced by a box with an integral sign within it. Even worse, a larger box adorned with an integral sign can act on a multitude of signals, a vector of inputs, to give an equal number of outputs. Figure 3-2 was an early example of this, while Fig. 9-8 will take it even further.

Yet another representation is the *signal flow graph* which will appear in Prelude 11, starting with Fig. P11-4. Here the boxes and symbols are left out altogether, and the transfer function is simply written alongside a line joining two nodes. The method is economic in graphic effort, but can disguise some alarming anomalies. Different representations are appropriate in different circumstances, but it is always important to be able to recognize the equivalence of the system they describe.

9-2 JUGGLING THE MATRICES

We start with the now-familiar matrix state equations

$$\dot{\mathbf{x}} = A\mathbf{x} + B\mathbf{u}$$

$$\mathbf{y} = C\mathbf{x}$$

A transformation of variables $\mathbf{w} = T\mathbf{x}$ changes these equations to

$$\dot{\mathbf{w}} = TAT^{-1}\mathbf{w} + TB\mathbf{u}$$

and
$$\mathbf{y} = CT^{-1}\mathbf{w}$$

In the prelude to this chapter we saw that a transformation T can usually be found which will make TAT^{-1} diagonal. How does this help us?

As soon as we have a diagonal system matrix, the system falls apart into a set of unconnected subsystems. Each equation for the derivative of one of the w's contains only that same w on the right-hand side. In the 'companion form' we saw that a train of off-axis ones defined each state variable to be the derivative of the one before, linking them all together (Sec. 8-6). There is no such linking in the diagonal case; each variable stands on its own. The only coupling between variables is their possible sharing of the inputs and their mixture to form the outputs (Fig. 9-1).

Each component of w represents one exponential function of time in the analytic solution of the differential equations. The equations are equivalent to

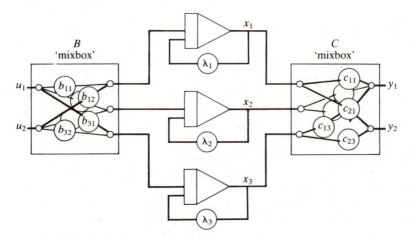

Figure 9-1 The B matrix applies a mixture of the inputs to the state integrators while the C matrix constructs the output from a mixture of states.

taking each transfer function expression, factorizing its denominator and splitting it into partial fractions.

When the system has a single input and a single output we can be even more specific. Such a system might be a filter, represented by one single transfer function. Now each of the subsystems is driven from the single input, and the outputs are mixed together to form the single output. The matrix B is an n-element column vector, while C is an n-element row vector. If we double the values of all the B elements and halve all those of C, the overall effect will be unchanged. In fact the gain through each term will be the product of a B coefficient and the corresponding C coefficient; only this product matters, so we can choose each of the B values to be unity and let C sort out the result. Equally we could let the C values be unity and rely on B to set the product (see Fig. 9-2).

In matrix terms, the inverse of the transformation matrix, T^{-1}, is obtained by stacking together the column vectors of the eigenvectors. Each of these vectors could be multiplied by a constant and it would still be an eigenvector. The transformation matrix is therefore far from unique, and can be fiddled to make B or C (in the single-input–single-output case) have elements that are all unity.

Let us bring the discussion back down to earth by looking at an example or two. Consider the second-order lag, or low-pass filter, defined by the transfer function

$$Y(s) = \frac{1}{(s + a)(s + b)} U(s)$$

The obvious way to split this into first-order subsystems is to regard it as

$$Y(s) = \frac{1}{s + a} \quad \frac{1}{s + b} U(s)$$

Applying the two lags one after the other suggests a state-variable representation

$$\dot{x}_1 = -bx_1 + u$$

$$\dot{x}_2 = -ax_2 + x_1$$

where

$$y = x_1$$

that is

$$\begin{bmatrix} \dot{x}_1 \\ \dot{x}_2 \end{bmatrix} = \begin{bmatrix} -b & 0 \\ 1 & -a \end{bmatrix} \begin{bmatrix} x_1 \\ x_2 \end{bmatrix} + \begin{bmatrix} 1 \\ 0 \end{bmatrix} u$$

$$y = \begin{bmatrix} 0 & 1 \end{bmatrix} \begin{bmatrix} x_1 \\ x_2 \end{bmatrix}$$

In analogue computer terms, this is shown in Fig. 9-3.

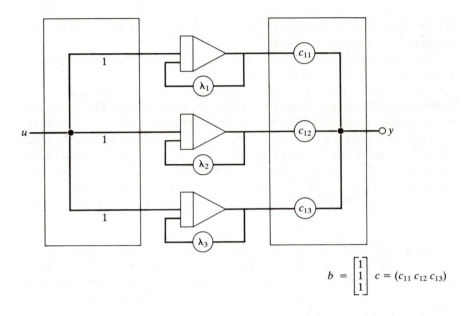

$$b = \begin{bmatrix} 1 \\ 1 \\ 1 \end{bmatrix} \quad c = (c_{11} \ c_{12} \ c_{13})$$

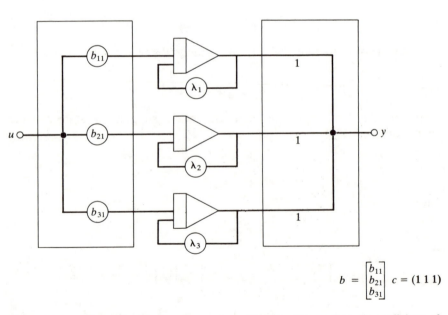

$$b = \begin{bmatrix} b_{11} \\ b_{21} \\ b_{31} \end{bmatrix} \quad c = (1 \ 1 \ 1)$$

Figure 9-2 In a single-input–single-output system with distinct eigenvalues, the coefficients of the B matrix can all be made unity or else the coefficients of C can all be unity.

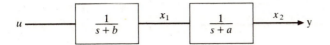

Figure 9-3 A system containing two cascaded lags.

Transforming the A matrix to diagonal form is equivalent to breaking the transfer function into partial fractions:

$$Y(s) = \frac{1}{b-a}\left(\frac{1}{s+a} - \frac{1}{s+b}\right)U(s)$$

This expression can be represented equally well by

$$Y(s) = W_1(s) + W_2(s)$$

where

$$W_1(s) = \frac{1}{b-a}\frac{1}{s+a}U(s)$$

$$W_2(s) = \frac{-1}{b-a}\frac{1}{s+b}U(s)$$

or by

$$Y(s) = \frac{1}{b-a}Z_1(s) - \frac{1}{b-a}Z_2(s)$$

where

$$Z_1(s) = \frac{1}{s+a}U(s)$$

$$Z_2(s) = \frac{1}{s+b}U(s)$$

In both cases the A matrix is diagonal, with non-zero elements $-a$ and $-b$. For the w's, however, the C matrix has both elements unity, while in the case of the z's it is the B matrix that is all unity.

The second-order lag response to a step input function has a zero initial derivative; the two first-order lags are mixed in such a way that their initial slopes are cancelled (see Fig. 9-4).

Exercise 9-2-1 *Write out the state equations of the above argument in full for the case where $a = 2$, $b = 3$.*

Exercise 9-2-2 *Write out the equations again for the case $a = b = 2$. What goes wrong?*

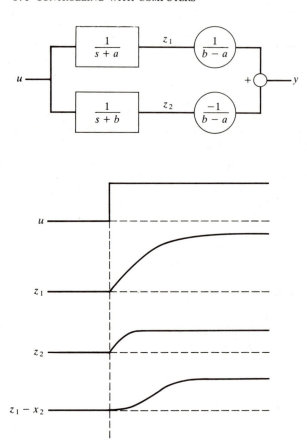

Figure 9-4 A second-order response as the difference of two first-order responses.

9-3 REPEATED ROOTS

There is always a snag. If all the roots are distinct, then the A matrix can be diagonalized using a matrix found from the eigenvectors and all is well. A repeated root throws a spanner in the works in the simplest of examples.

In Sec. 9-2 we saw that a second-order step response could be derived from the difference between two first-order responses. However, if the time constants coincide the difference becomes zero, and hence useless. The analytic solution contains not just a term $\exp(-at)$ but another term $t\exp(-at)$; the exponential is multiplied by time.

We simply have to recognize that the two cascaded lags can no longer be separated, but must be simulated in that same form. If there are three equal roots, then there may have to be three equal lags in cascade. Instead of

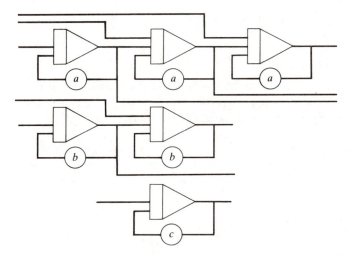

Figure 9-5 A form of the simulation when there are repeated roots.

achieving a diagonal form, we may only be able to reduce the A matrix to a form such as

$$
\begin{bmatrix}
a & 1 & 0 & 0 & 0 & 0 \\
0 & a & 1 & 0 & 0 & 0 \\
0 & 0 & a & 0 & 0 & 0 \\
0 & 0 & 0 & b & 1 & 0 \\
0 & 0 & 0 & 0 & b & 0 \\
0 & 0 & 0 & 0 & 0 & c
\end{bmatrix}
$$

This, the *Jordan canonical form*, represents a simulation of the form shown in Fig. 9-5.

Repeated roots do not always mean that a diagonal form is impossible. Two completely separate single lags, each with their own input and output, can be combined into a single set of system equations. The A matrix is then of course diagonal, since there is no reaction between the two subsystems. If the system is single-input–single-output, however, repeated roots always mean trouble.

In another case, the diagonal form is possible but not the most desirable. Suppose that some of the roots of the characteristic equation are complex. It is not easy to apply a feedback gain of $2 + 3j$ around an integrator! The second-order equation with roots $-k \pm jn$,

$$ \ddot{y} + 2k\dot{y} + (n^2 + k^2)y = u $$

is more neatly represented for simulation by

$$
\begin{bmatrix} \dot{x}_1 \\ \dot{x}_2 \end{bmatrix} \begin{bmatrix} -k & n \\ -n & -k \end{bmatrix} \begin{bmatrix} x_1 \\ x_2 \end{bmatrix} + \begin{bmatrix} 0 \\ 1 \end{bmatrix} u \tag{9-1}
$$

than by a set of matrices with complex coefficients. If we accept quadratic terms as well as linear factors, any polynomial with real coefficients can be factorized without having to resort to complex numbers.

Exercise 9-3-1 *Derive state equations for the system $\ddot{y} + y = u$. Find the Jordan canonical form, and also find a form in which a real simulation is possible, similar to the example of expression (9-1). Sketch the simulation.*

9-4 CONTROLLABILITY AND OBSERVABILITY

With the system equation reduced to diagonal form, we saw the w state variables standing alone, with no cross-coupling between them. The only signals that can affect their behaviour are the inputs. Suppose, however, that one or more of these state variables has no input connected to it. Then there is no way in which those variables can be controlled. Such variables are *uncontrollable.*

If the poles associated with all the uncontrollable variables are stable, then the system performance as a whole can still be acceptable. No amount of feedback can move the positions of these poles, however.

Another alarming possibility is that there are state variables that affect none of the outputs. Their behaviour is not measured in any way and so feedback will again have no effect on their pole values. These variables are *unobservable.* You may say that since the variables do not affect the outputs, you do not care what they do. Unfortunately the outputs for control purposes are the transducers and sensors of the control system; the system could well be an aircraft carrying passengers who have sensors of their own, which are making them turn somewhat green.

Some of the variables could be both uncontrollable and unobservable, that is to say, they might have neither input nor output connections. The system can now be partitioned into four subsystems (Fig. 9-6):

1. Controllable and observable
2. Controllable but not observable
3. Observable but not controllable
4. Neither controllable nor observable

Clearly any system model built to represent a transfer function must be both controllable and observable—there is no point in adding uncontrollable or unobservable modes. The control engineer is baffled by the fourth type of

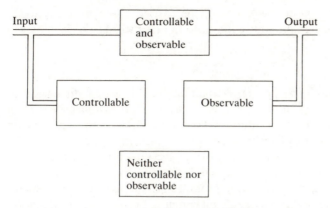

Figure 9-6 A system can be divided into four subsystems (not all of which need to exist).

subset, somewhat frustrated by sets two and three, and is really only at home in applying control to the first.

Note that a system can be controllable and observable with very few inputs and outputs. In the two-cascaded-lags example of Sec. 9-2, we saw that the single input was applied to the first lag, while the single output was taken from the second. So how can we be sure that we can control the state corresponding to the second lag or observe that of the first? When the system is represented in diagonal form all the necessary input and output coefficients are non-zero to ensure controllability and observability. Can we deduce these properties from the general matrix state equations without going to the trouble of transforming them?

Let us first set all the state variables to zero in the system

$$\dot{\mathbf{x}} = A\mathbf{x} + B\mathbf{u}$$

If we apply a brief input $\mathbf{u}$ for a time δt, then to a first approximation the state will be carried to the value $B\mathbf{u} \, \delta t$. We can set just the first component of the input to a non-zero value, in which case the state will move in the vector direction of the first column of B. If there is a second input, we can make that non-zero instead, perturbing the state in the direction of the second column of B. In this way we see that the columns of B form a set of vectors in whose directions we can drive the state vector, with immediate effect on its velocity.

Now we must refresh our memory of the concept of rank. If B has three columns which are linearly independent, then its rank is three. If the system has only three state variables, then our troubles are at an end; the system is controllable in a particularly simple way, since we can immediately impose any chosen velocity onto any selected state. However, suppose there is only one input, so that B can only have a rank of one. What then?

Having applied a brief input $u_1 = 1$, we see the state carried to a value $\mathbf{b}_1 \, \delta t$. Here $\mathbf{b}_1$ represents the first (and in this case only) column of the matrix

B. If we remove the input, the state will be carried along by the velocity imposed by the *A* term in the state equations. In other words, the velocity will be determined by

$$\dot{x} = A\mathbf{b}_1 \, \delta t$$

This velocity might be in a new direction, adding a new dimension to those in which the state can be driven. We must consider not just the rank of *B* alone but of *B* taken with the additional columns of *AB*. However, if we can manoeuvre the inputs to take the state in the direction of a column of *AB*, then by waiting a little we will see it move in the direction of *AAB*.

It looks at first as though we will have to investigate an infinite succession of matrices, but the Cayley–Hamilton theory comes to the rescue. It can be deduced that if *A* is 3 by 3, then it can be expressed as a combination of the matrices *I*, *A* and A^2. To test for controllability of a third-order system, therefore, we need only examine the rank of the matrix formed by stacking together the columns

$$(B \,|\, AB \,|\, A^2 B)$$

For an *n*th order system we must of course take terms up to $A^{n-1}B$.

It is time for some exercises.

Exercise 9-4-1 *Is the following system controllable?*

$$\dot{x} = \begin{bmatrix} 0 & 1 \\ -1 & -2 \end{bmatrix} x + \begin{bmatrix} 0 \\ 1 \end{bmatrix} u$$

AB is $(1 \quad -2)'$, so to assess controllability we must examine the rank of

$$(B \,|\, AB) = \begin{bmatrix} 0 & 1 \\ 1 & -2 \end{bmatrix}$$

which is clearly two, since its determinant is non-zero (value -1).

Exercise 9-4-2 *Is this system also controllable?*

$$\dot{x} = \begin{bmatrix} 0 & 1 \\ -1 & -2 \end{bmatrix} x + \begin{bmatrix} -1 \\ 1 \end{bmatrix} u$$

Exercise 9-4-3 *Is this third system controllable?*

$$\dot{x} = \begin{bmatrix} 0 & 1 \\ -1 & -2 \end{bmatrix} x + \begin{bmatrix} 1 \\ 0 \end{bmatrix} u$$

Exercise 9-4-4 *Transform the above systems to Jordan canonical form to see the structure of their control from a simulation diagram.*

This is worth working through here in the text, since there are some interesting lessons to be drawn.

Taking the determinant of $A - \lambda I$, we see that there are two equal roots of value 1. How can we find two eigenvectors to build the transformation? We cannot.

In general, if we call the eigenvectors $\mathbf{p}_1$, etc., then we will be exploiting the relationship

$$A(\mathbf{p}_1, \mathbf{p}_2, \ldots) = (\lambda_1 \mathbf{p}_1, \lambda_2 \mathbf{p}_2, \ldots)$$

$$= \Lambda(\mathbf{p}_1, \mathbf{p}_2, \ldots)$$

to find the diagonalizing transformation. If Λ is truly diagonal, then the second term, for example, would be

$$A\mathbf{p}_2 = \lambda_2 \mathbf{p}_2$$

and its solution would yield an eigenvector. When we accept that Λ cannot be diagonal, but must have an extra 1 in the row-1 column-2 position, the equation acquires an extra term to become

$$A\mathbf{p}_2 = \mathbf{p}_1 + \lambda_2 \mathbf{p}_2$$

$\mathbf{p}_2$, no longer an eigenvector, is the second column of the matrix defining the transformation. We see that within a block corresponding to repeated roots, each vector except the first is given by an equation involving the preceding vector:

$$(A - \lambda_r I)\mathbf{p}_r = \mathbf{p}_{r-1}$$

Turning our attention back to the example, we see that if we substitute $\lambda = -1$ we get

$$A - \lambda I = \begin{bmatrix} 1 & 1 \\ -1 & -1 \end{bmatrix}$$

Now $(A - \lambda I)\mathbf{p}_1 = 0$ gives us the first column of the transformation, the eigenvector $(1, \ -1)'$. For the second column we have

$$(A - \lambda I)\mathbf{p}_2 = \mathbf{p}_1$$

that is

$$\begin{bmatrix} 1 & 1 \\ -1 & 1 \end{bmatrix}\begin{bmatrix} x_1 \\ x_2 \end{bmatrix} = \begin{bmatrix} 1 \\ -1 \end{bmatrix}$$

with solution $(1 \quad 0)'$. The transformation is defined by

$$T^{-1} = \begin{bmatrix} 1 & 1 \\ -1 & 0 \end{bmatrix} .$$

so

$$T = \begin{bmatrix} 0 & -1 \\ 1 & 1 \end{bmatrix}$$

After its transformation, the A matrix becomes

$$\begin{bmatrix} -1 & 1 \\ 0 & -1 \end{bmatrix}$$

In Exercise 9-4-1, the new B matrix, TB, becomes

$$\begin{bmatrix} 0 & -1 \\ 1 & 1 \end{bmatrix} \begin{bmatrix} 0 \\ 1 \end{bmatrix} = \begin{bmatrix} -1 \\ 1 \end{bmatrix}$$

which has both elements non-zero. In Exercise 9-4-2, we have for TB:

$$\begin{bmatrix} 0 & -1 \\ 1 & 1 \end{bmatrix} \begin{bmatrix} 1 \\ -1 \end{bmatrix} = \begin{bmatrix} 1 \\ 0 \end{bmatrix}$$

and so the second state variable cannot be excited. This system is uncontrollable, as you will already have found by looking at the rank of $(B \mid AB)$. In the third example, we have for TB:

$$\begin{bmatrix} 0 & -1 \\ 1 & 1 \end{bmatrix} \begin{bmatrix} 1 \\ 0 \end{bmatrix} = \begin{bmatrix} 0 \\ 1 \end{bmatrix}$$

Despite appearances, this system is controllable. The second state variable responds to the input, and in turn influences the first state. All three systems are illustrated in Fig. 9-7.

Can we deduce observability in a similarly simple way by looking at the rank of a set of matrices?

The equation

$$\mathbf{y} = C\mathbf{x}$$

is likely to represent fewer outputs than states; we have not enough simultaneous equations to solve for $\mathbf{x}$. If we ignore all problems of noise and continuity, we can consider differentiating the outputs to obtain

$$\dot{\mathbf{y}} = C\dot{\mathbf{x}}$$
$$= C(A\mathbf{x} + B\mathbf{u})$$

that is

$$\dot{\mathbf{y}} - CB\mathbf{u} = CA\mathbf{x}$$

Now we can look among the rows of A and CA to find enough independent simultaneous equations to solve for the state; of course we need as many independent rows as there are state variables.

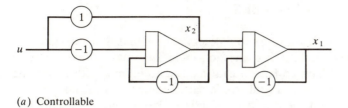

(*a*) Controllable

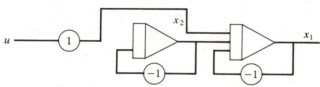

(*b*) x_2 is not controllable

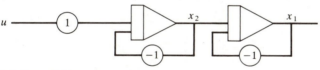

(*c*) Controllable

Figure 9-7 Simulations of (*a*) Exercise 9-4-1, (*b*) Exercise 9-4-2, (*c*) Exercise 9-4-3.

If we still are not satisfied, we can differentiate again, giving us some more equations involving CA^2, and so on. Once more we use the Cayley–Hamilton theory to call a halt to the search and deduce that the system is observable if we can find n independent rows among those of

$$\begin{bmatrix} C \\ \overline{CA} \\ \overline{CA^2} \\ \cdots \\ \cdots \\ CA^{n-1} \end{bmatrix}$$

In other words, the rank of the matrix above must be n.

9-5 PRACTICAL OBSERVERS

It is unwise, to say the least, to try to differentiate a signal. Some devices that claim to be differentiators are in fact mere high-pass filters. A true differentiator would have to have a gain tending to infinity with increasing frequency. Any noise in the signal would cause immense problems.

Let us forget about the problems of differentiation and instead address the direct problem of deducing the state of a system from its outputs.

First we have to assume that we have a complete knowledge of the state equations. Can we not then set up a simulation of the system, and by applying the same inputs simply measure the states of the model? This might succeed if the system has only poles that represent rapid settling—the sort of system that does not really need feedback control! Suppose instead that the system is a motor-driven position controller. The output involves at least one pure integration of the input signal; any error in setting up the initial conditions of the model will persist indefinitely.

Let us not give up the idea of a model completely. Suppose that there are indeed some measurements of the system output, enough to represent observability. Can we not use these signals to 'pull the model into line' with the state of the system? This is the principle underlying the *Kalman filter*.

The system, as usual, is given by

$$\dot{\mathbf{x}} = A\dot{\mathbf{x}} + B\mathbf{u} \qquad (9\text{-}2)$$

$$\mathbf{y} = C\mathbf{x} \qquad (9\text{-}3)$$

We can set up a simulation of the system, having similar equations, where the variables are $\hat{x}$ and $\hat{y}$; the 'hats' mark the variables as estimates. In the real system, we can only influence the variables through the input and the matrix B. In the model, we can 'cheat' and apply an input signal directly to any integrator and hence to any state variable that we choose. We can, for instance, calculate the error between the measured outputs $\mathbf{y}$ and the estimated outputs $\hat{y}$ given by $C\hat{x}$, and mix this signal among the state integrators in any way we wish. The model equations then become

$$\hat{\mathbf{x}} = A\hat{\mathbf{x}} + B\mathbf{u} + K(\mathbf{y} - C\hat{\mathbf{x}}) \qquad (9\text{-}4)$$

$$\hat{\mathbf{y}} = C\hat{\mathbf{x}} \qquad (9\text{-}5)$$

The corresponding system is illustrated in Fig. 9-8.

Clearly the model states have two sets of inputs now, one corresponding to the system input and the other taken from the system output. The model 'A matrix' has also been changed, as we see by rewriting Eq. (9-4) to obtain

$$\dot{\hat{\mathbf{x}}} = (A - KC)\hat{\mathbf{x}} + B\mathbf{u} + K\mathbf{y} \qquad (9\text{-}6)$$

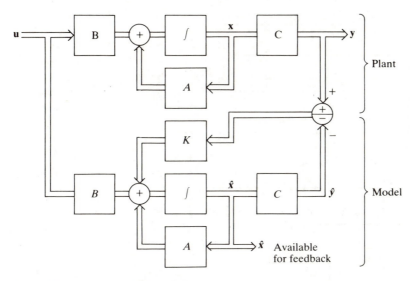

Figure 9-8 Structure of a Kalman filter.

To see just how well we might succeed in tracking the system state variables, we can combine Eqs (9-2) to (9-5) to give differential equations for the estimation error:

$$\dot{\hat{x}} - \dot{x} = A(\hat{x} - x) + KC(\hat{x} - x)$$
$$= (A - KC)(\hat{x} - x)$$

The eigenvalues of $A - KC$ will determine how rapidly the model states settle down to mimic the states of the plant. These are the roots of the model, as defined in Eq. (9-6). If the system is observable, we should be able to choose the coefficients of K to place the roots wherever we wish; the choice will be influenced by the noise levels we expect to find on the signals.

Exercise 9-5-1 *A motor is described by two integrations, from input drive to output position. The velocity is not directly measured. We wish to achieve a well-damped position control and so need a velocity term to add. Design an observer.*

The system equations for this exercise may be written

$$\dot{x} = \begin{bmatrix} 0 & 1 \\ 0 & 0 \end{bmatrix} x + \begin{bmatrix} 0 \\ 1 \end{bmatrix} u$$

$$y = \begin{bmatrix} 1 & 0 \end{bmatrix} x$$

(9-7)

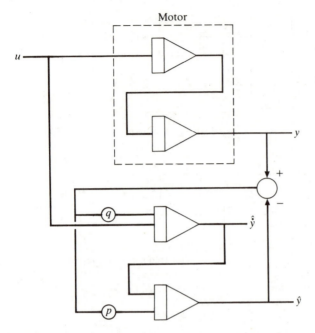

Figure 9-9 Kalman filter solution to Exercise 9-5-1.

We can choose the matrix K to be any $(p \quad q)'$, and so we arrive at

$$A - KC = \begin{bmatrix} 0 & 1 \\ 0 & 0 \end{bmatrix} - \begin{bmatrix} p \\ q \end{bmatrix} \begin{bmatrix} 1 & 0 \end{bmatrix}$$

$$= \begin{bmatrix} -p & 1 \\ -q & 0 \end{bmatrix}$$

The eigenvalues of this are given by $\det(A - KC - \lambda I) = 0$, that is

$$\lambda^2 + p\lambda + q = 0$$

We can place the roots wherever we wish, by suitable choice of p and q (see Fig. 9-9).

It looks as though we now have both position and velocity signals to feed around our motor system. If this is really the case, then we can put the closed loop poles wherever we wish. It seems that anything is possible—keep hoping.

Exercise 9-5-2 *Design observer and feedback for the motor of Exercise 9-5-1 to give a response characteristic having two equal roots of 0.1 seconds, and an observer error characteristic having equal roots of 0.5 seconds. Sketch the corresponding circuit containing the integrators.*

9-6 REDUCED-STATE OBSERVERS

In the last exercise, we seemed to have to use a second-order observer to deduce a single state variable. Is there a more economic way to make an observer? Luenberger suggested the answer.

Suppose that we do not wish to estimate all the components of the state, but only a selection, given by $S\mathbf{x}$. We would like to set up a modelling system having states $\mathbf{z}$, and taking inputs $\mathbf{u}$ and $\mathbf{y}$, such that the values of the $\mathbf{z}$ components tend to the signals we wish to observe. This appears less complicated when written algebraically! The observer equations are

$$\dot{\mathbf{z}} = P\mathbf{z} + Q\mathbf{u} + R\mathbf{y}$$

and we want all the components of $\mathbf{z} - S\mathbf{x}$ to tend to zero in some satisfactory way.

Now we see that the derivative of $\mathbf{z} - S\mathbf{x}$ is given by

$$\dot{\mathbf{z}} - S\dot{\mathbf{x}} = P\mathbf{z} + Q\mathbf{u} + R\mathbf{y} - S(A\mathbf{x} + B\mathbf{u})$$

$$= P\mathbf{z} + Q\mathbf{u} + RC\mathbf{x} - S(A\mathbf{x} + B\mathbf{u})$$

$$= P\mathbf{z} + (RC - SA)\mathbf{x} + (Q - SB)\mathbf{u}$$

We would like this to reduce to

$$\dot{\mathbf{z}} - S\dot{\mathbf{x}} = P(\mathbf{z} - S\mathbf{x})$$

where P represents a system matrix giving rapid settling to zero. For this to be the case,

$$-PS = RC - SA$$

and $$Q - SB = 0$$

that is

$$Q = SB \tag{9-8}$$

and $$RC = SA - PS \tag{9-9}$$

Exercise 9-6-1 *Design a first-order observer for the motor problem of Exercise 9-5-1.*

Exercise 9-6-2 *Apply the observed velocity to achieve closed loop control as specified in Exercise E9-5-2.*

The first problem hits an unexpected snag, as you will see.

If we refer back to the system state equations of (9-7), we see that

$$A = \begin{bmatrix} 0 & 1 \\ 0 & 0 \end{bmatrix} \qquad B = \begin{bmatrix} 0 \\ 1 \end{bmatrix} \qquad C = \begin{bmatrix} 1 & 0 \end{bmatrix}$$

If we are only interested in estimating the velocity, then we have

$$S = [0 \quad 1]$$

Now $Q = SB$, which has a scalar value of 1. P becomes a single parameter defining the speed with which z settles, and we may set

$$P = -k$$

We see that

$$SA - PS = (0 \quad 0) - (-k)(0 \quad 1)$$

$$= (0 \quad k)$$

We now need R, here a simple constant r, such that

$$SA - PS = RC$$

that is

$$(0 \quad k) = r(1 \quad 0)$$

Clearly there is no possible solution, apart from making r and k zero.

Do not give up! We actually need the velocity so that we can add it to a position term for feedback. Suppose that instead of pure velocity we try for a mixture of position plus a times velocity. Then

$$Sx = (1 \quad a) \, x$$

This time

$$SA - PS = (0 \quad 1) - (-k)(1 \quad a)$$

$$= (1 \quad 1 + ka)$$

Now we can equate this to RC if $r = 1$ and $1 + ka = 0$. From this last equation it looks as though a will be negative, with the value $-1/k$. To construct the velocity we will have to take k times $(y - z)$. We notice that now Q has the value $-1/k$. Mission accomplished!

We can set out the observer clearly as

$$\dot{z} = -kz - \frac{1}{k} u + y$$

and

$$\hat{y} = k(y - z)$$

To see it in action, let us attack Exercise 9-6-2. We require a response similar to that specified in Exercise 9-5-2. We want the closed loop response to have equal roots of 0.1 seconds, with a single observer settling time constant of 0.5 seconds.

For the 0.5 second observer time constant, we make $k = 2$.

For the feedback we now have all the states (or their estimates) available, so instead of considering $A + BFC$ we only have $A + BF$ to worry about. To achieve the required closed loop response we can propose algebraic values for the feedback parameters in F, substitute them to obtain $A + BF$ and then from the eigenvalue determinant derive the characteristic equation. Finally, we equate coefficients between the characteristic equation and the equation with the roots we are trying to fiddle. Sounds complicated.

In this simple example we can look at its system equations in the 'traditional form':

$$\ddot{y} = u$$

The time constants of 0.1 seconds imply root values of 10, so when the loop is closed the behaviour of y must be described by

$$\ddot{y} + 20\dot{y} + 100y = 0$$

that is

$$\ddot{y} = -100y - 20\dot{y}$$

so we must set the system input to

$$u = -100y - 20\hat{\dot{y}}$$
$$= -100y - 20 \times 2(y - z)$$
$$= -140y + 40z$$

where z is given by an added subsystem

$$\dot{z} = -2z - 0.5u + y$$

Before leaving the problem, let us look at the observer in transfer function terms. We can write

$$(s + 2)Z = -0.5U + Y$$

so the equation for the input u becomes

$$U = -140Y + 40Z$$

$$= -140Y + 40\,\frac{1}{s + 2}(-0.5U + Y)$$

So, tidying up, we have

$$(s + 2)U + 20U = -140(s + 2)Y + 40Y$$
$$(s + 22)U = -(140s + 60)Y$$

$$U = -\frac{140s + 60}{s + 22}Y$$

The whole process whereby we calculate the feedback input u from the position output y has boiled down to nothing more than a simple phase advance! To make matters worse, it is not even a very practical phase advance, since it requires the high frequency gain to be over fifty times that at low frequency; we can certainly expect noise problems from it.

Do not take this as an attempt to belittle the observer method. Over the years, engineers have developed intuitive techniques to deal with common problems, and only those techniques that were successful have survived. The fact that phase advance can be shown to be equivalent to the application of an observer detracts from neither method—just the reverse.

By tackling a problem systematically, analysing general linear feedback of states and estimated states, the whole spectrum of solutions can be surveyed to make a choice. The reason that the solution to the last problem was impractical had nothing to do with the method, but depended entirely on our arbitrary choice of settling times for the observer and for the closed loop system. We should instead have left the observer time constant as a parameter for later choice, and we could then have imposed some final limitation on the ratio of gains of the resulting phase advance. Through this method we would at least know the implications of each choice.

If the analytic method does have a drawback, it is that it is too powerful. The engineer is presented with a wealth of possible solutions, and is left agonizing over the new problem of how to limit his choice to a single answer. Some design methods are tailored to reduce these choices; as often as not, they throw the baby out with the bathwater. *Diadic feedback* suggests that the inputs can be driven with varying amounts of the same mixture of the outputs, for example. It is highly successful in reducing the number of parameters to choose, still placing the poles as directed, but the zeros and hence the system responses to the various inputs are a very different matter.

Let us go on to examine linear control in its most general terms.

9-7 CONTROL WITH ADDED DYNAMICS

We can scatter dynamic filters all around the control system, as shown in Fig. 9-10. The signals shown in Fig. 9-10 can have many components; the blocks represent transfer function matrices, not just simple transfer functions. This means that we have to use caution when applying the methods of *block diagram manipulation*.

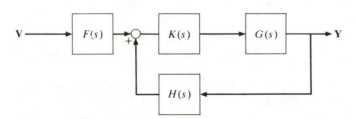

Figure 9-10 Feedback around $G(s)$ with three added filters.

In the case of scalar transfer functions, we can unravel complicated structures by the relationships illustrated in Fig. 9-11.

However complicated our controller may be, it simply takes inputs from the 'command input' **v** and the system output **y** and delivers a signal to the system input **u**. It can be described by just two transfer function matrices. We have one transfer function matrix linking the command input to the system input, which we will call the *feedforward matrix*. We have another matrix linking the output back to the system input, not surprisingly called the *feedback matrix*.

If the controller has internal state **z**, then we can append these components to the system state **x** and write down a set of state equations for our new, bigger system. Suppose that we started with

$$\dot{\mathbf{x}} = A\mathbf{x} + B\mathbf{u}$$

$$\mathbf{y} = C\mathbf{x}$$

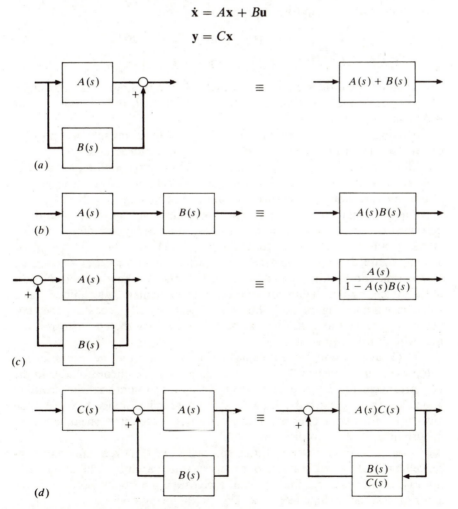

Figure 9-11 Some rules of block diagram manipulation.

and added a controller with state **z**, where

$$\dot{z} = Kz + Ly + Mv$$

Now we apply signals from the controller dynamics, the command input and the system output to the system input **u**:

$$u = Fy + Gz + Hv$$

so that

$$\dot{x} = (A + BFC)x + BGz + BHv$$

and

$$\dot{z} = LCx + Kz + Mv$$

We end up with a composite matrix state equation

$$\begin{bmatrix} \dot{x} \\ \dot{z} \end{bmatrix} = \begin{bmatrix} A + BFC & | & BG \\ \hline LC & | & K \end{bmatrix} \begin{bmatrix} x \\ z \end{bmatrix} + \begin{bmatrix} BH \\ M \end{bmatrix} v$$

We could even consider a new 'super output' by mixing **y**, **z** and **v** together, but with the coefficients of all the matrices F, G, H, K, L and M to choose, life is difficult enough as it is.

Moving back to the transfer function form of the controller, is it possible or even sensible to try feedforward control alone? Indeed it is.

Suppose that the system is a simple lag, slow to respond to a change of demand. It makes sense to apply a large initial change of input, to get the output moving, and then to turn the input back to some steady value.

Suppose that we have a simple lag with time constant 5 seconds, described by the transfer function $1/(1 + 5s)$. If we apply a feedforward filter at the input, having transfer function $(1 + 5s)/(1 + s)$ (here is that phase advance again!), then the overall response will have the form $1/(1 + s)$—the 5 second time constant has been reduced to 1 second. We have cancelled a pole of the system by putting an equal zero in the controller—a technique called *pole cancellation*. Figure 9-12 shows the effect. Strictly speaking, the time functions $v(t)$, $u(t)$ and $y(t)$ should be represented by their transforms, but that would not help clarity.

Take care. Although the pole has been removed from the response, it is still present in the system. Moreover, it has been made uncontrollable in the new arrangement. If it is benign, representing a transient that dies away in good time, then we can bid it farewell. If it is close to instability, however, then any error in the controller parameters or any unaccounted 'sneak' input or disturbance can lead to disaster.

If we put on our observer-tinted spectacles, we can even represent pole-cancelling feedforward in terms of observed state variables. The observer in this case is rather strange in that **z** can receive no input from **y**—otherwise the controller would include feedback. Try the following exercise.

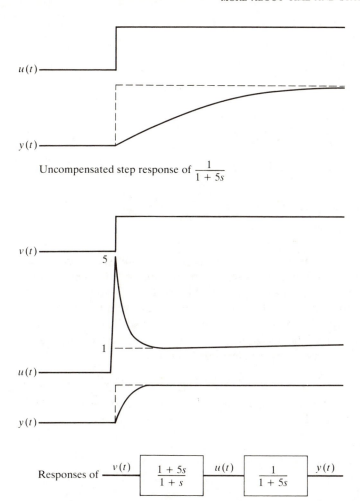

Figure 9-12 Responses with feedforward pole cancellation.

Exercise 9-7-1 *Set up the state equations for a 5 second lag. Add an estimator to estimate the state from the input alone. Feed back the estimated state to obtain a response with a 1 second time constant.*

9-8 AN UNUSUAL EXAMPLE OF AN INVERTED PENDULUM

The inverted pendulum is a favourite target for virtuoso controllers, and appears in several forms. The more difficult version is analogous to balancing an inverted broom on one's hand, while simpler tasks merely involve position

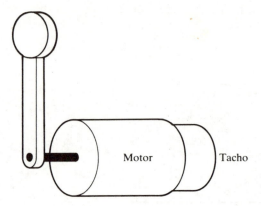

Figure 9-13 Inverted pendulum.

and velocity feedback in a system disturbed by an inverted torque. The example described here is simple to set up and has an amusing performance.

The pendulum to be balanced is attached to the horizontal shaft of a servomotor, as shown in Fig. 9-13. The servomotor carries a tachogenerator, so that a velocity signal is available, but there is no position feedback. When the motor is unpowered, the pendulum will hang vertically downwards and if slightly disturbed will perform oscillations described by

$$\ddot{y} = -ky$$

If the pendulum is lifted to the top of its arc, it is of course unstable. Its motion is now described by

$$\ddot{y} = +ky$$

When we apply a drive signal to the motor, we will have

$$\ddot{y} = ky + au$$

Can we close a loop using velocity feedback alone to drive the pendulum to the top of its arc? If we can, the pendulum will seek the upward vertical however the body of the motor is turned.

We must, of course, allow dynamics within the feedback loop, but we can achieve our purpose with just a single extra state, i.e. with one integrator. We can apply the tacho signal and local feedback to this integrator so that its state, z, is described by

$$\dot{z} = b\dot{y} + cz$$

while u can be described as a mixture of $\dot{y}$ and z as

$$au = d\dot{y} + ez$$

that is

$$\ddot{y} = ky + d\dot{y} + ez$$

Let us put these equations into state form. If we choose state variables y, $\dot{y}$ and z then we have

$$\dot{\mathbf{x}} = \begin{bmatrix} 0 & 1 & 0 \\ k & d & e \\ 0 & b & c \end{bmatrix} \mathbf{x}$$

When we look for the eigenvalues and evaluate the characteristic polynomial, $\det(\lambda I - A)$, we find

$$\lambda^3 + \lambda^2(-c - d) + \lambda(cd - eb - k) + kc = 0$$

Now k is positive, a property of the pendulum, but we can choose b, c, d and e at will. Indeed, we can choose the coefficients to achieve any placement of the poles we wish! First note that since the constant term must be positive we can deduce that c is also positive; the controller represents an unstable system in itself.

When we perform the necessary algebraic steps to represent the controller as a transfer function, we find that it reduces to

$$u = -f \frac{s + g}{s - c} \dot{y}$$

where f, g and c are all positive, while the pendulum relates $\dot{y}$ to u by

$$\dot{y} = a \frac{s}{s^2 - k} u$$

I can assure you that the controller does really work (see Fig. 9-14). The friction that damps out the oscillations when the inert pendulum hangs downwards will now cause the pendulum to perform small oscillations about the upward vertical.

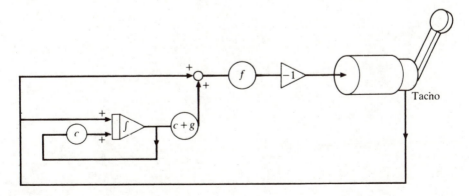

Figure 9-14 Control schematic.

9-9 CONCLUSION

Despite all the sweeping generalities. the subject of applying control to a system is far from closed. All these discussions have assumed that both system and controller will be linear; we saw back in Chapter 3 that considerations of signal and drive limiting can be essential. We have also only looked at feedback in continuous time, using analogue control, when we are eager to put digital computers to use. Through an uncritical eye the efforts of computers can appear continuous, but as soon as the task becomes complex and the response speeds become swift, it is all too obvious that the computer operates in fits and starts.

In the next chapter, we must look at the problems of control in *discrete time*, when control can only be applied at predetermined intervals.

DIFFERENCE OPERATORS AND FINITE TIME APPROXIMATIONS

P10-1 INTRODUCTION

The philosophers of Ancient Greece found the concept of continuous time quite baffling, getting tied up in paradoxes such as that of Xeno. Today's students are brought up on a diet of integral and differential calculus, taking velocities and accelerations in their stride, so that it is finite differences which may appear quite alien to them.

A few fundamental definitions have to be established, after which it will appear clear that discrete time systems can be managed more expediently than continuous ones. It is when the two are mixed that the control engineer must be wary.

This prelude clears the way for the next chapter, in which we can really get to grips with the theory of computer control.

P10-2 FINITE DIFFERENCES—THE BETA OPERATOR

We built our concepts of continuous control on the differential operator, *rate of change*. We let time change by a small amount δt, resulting in a change of $x(t)$ given by

$$\delta x = x(t + \delta t) - x(t)$$

Then we examined the limit of the ratio $\delta x/\delta t$ as we made δt tend to zero.

When we let a computer in on the act the rules must change. $x(t)$ is measured not continuously but at some sequence of times with fixed intervals. We might know $x(0)$, $x(0.01)$, $x(0.02)$ and so on; between these values the function is a mystery. The idea of letting δt tend to zero is useless; the only sensible value for δt is equal to the sampling interval. It makes little sense to go on labelling the functions with their time value, $x(t)$. We might as well acknowledge them to be a sequence of discrete samples and label them $x(n)$ according to their sample number.

We must be satisfied with an approximate sort of derivative, where δt takes a fixed value of one sampling interval, τ. We are left with another subtle problem which is important nonetheless. Should we take our difference as 'next value minus this one' or as 'this value minus last one'? If we could attribute the difference to lie at a time midway between the samples there would be no such question, but that does not help when we tag variables with 'sample number' rather than time.

The point of the system equations, continuous or discrete time, is to be able to predict the future state from the present state and the input. If we have the equivalent of an integrator, we might write

$$f(n + 1) = f(n) + \tau g(n)$$

which settles the question in favour of the forward difference

$$g(n) = \frac{f(n + 1) - f(n)}{\tau}$$

It looks as though we can only calculate $g(n)$ from the f sequence if we already know a future value, $f(n + 1)$. In the continuous case, however, we always dealt with integrators and would not consider differentiating a signal, so perhaps this will not prove a disadvantage.

We might define this approximation to differentiation as an operator, β, involving a 'time advance' operation. We could now write

$$g(n) = \beta(f(n))$$

where we might previously have written

$$g(t) = D(f(t))$$

with a differential operator, D. The inverse of the D operator is the integrator, at the heart of all continuous simulation. The inverse of the β operator is the crude numerical process of Euler integration:

$$x(n + 1) = x(n) + \tau \dot{x}(n)$$

The second-order differential equation

$$\ddot{x} + a\dot{x} + bx = u$$

might be approximated in discrete time by

$$\beta^2 x + a\beta x + bx = u$$

Will we look for stability in the same way as before? You will remember that we looked at the roots of the quadratic

$$m^2 + am + b = 0$$

and were concerned that all the roots should have negative real parts. This concern was originally derived from the fact that $\exp(mt)$ was an eigenfunction of a linear differential equation—that if an input $\exp(mt)$ was applied, the output would be proportional to the same function. Is this same property true of the finite difference operator β?

We can calculate

$$\beta(\exp(mt)) = \frac{\exp[m(t + \tau)] - \exp(mt)}{\tau}$$

$$= \exp(mt)\frac{\exp(m\tau) - 1}{\tau}$$

which is certainly proportional to $\exp(mt)$. We would therefore be wise to look for the roots of the characteristic equation, just as before, and plot them in the complex plane.

Just a minute, though. Will the imaginary axis still be the boundary between stable and unstable systems? In the continuous case, we found that a sine wave, $\exp(j\omega t)$, emerged from the D operator with an extra multiplying factor $j\omega$, represented by a point on the imaginary axis. If we subject the sine wave to the β operator we get

$$\beta(\exp(j\omega t)) = \exp(j\omega t)\frac{\exp(j\omega\tau) - 1}{\tau}$$

As ω varies, the gain moves along the locus defined by

$$\text{real part} = \frac{\cos(j\omega\tau) - 1}{\tau}$$

$$\text{Imaginary part} = \frac{\sin(j\omega\tau)}{\tau}$$

This is a circle, centre $-1/\tau$ and radius $1/\tau$, which touches the imaginary axis at the origin but curves round again to cut the real axis at $-2/\tau$. It is shown in Fig. P10-1. It is not hard to show that for stability the roots of the characteristic equation must lie inside this circle.

The system

$$\dot{x} = -ax$$

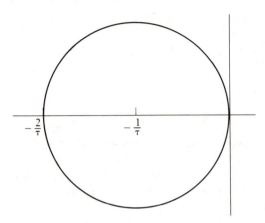

Figure P10-1 β-plane stability circle.

has a pole at $-a$ and is stable no matter how large a is made. On the other hand, the discrete time approximation to this system

$$\beta x = -ax$$

also has a pole at $-a$, but if a is increased to $2/\tau$ the pole emerges from the left margin of the stability circle and oscillation results. This is the step length problem in disguise.

If we take a set of poles in the s plane and approximate them by poles in the same position in the β plane, they must be considerably nearer the origin than $1/\tau$ for the approximation to ring true. If we want a near-perfect representation, we must map the s poles to β poles in somewhat different locations. Then again, the β plane might not give the clearest impression of what is happening.

P10-3 The z TRANSFORM

In the β operator, we looked for a discrete time approximation to differentiation. Why? Because we had some differential equations which we could solve to predict the future behaviour of the system. When we approximate

$$\dot{x} = -ax$$

by

$$\beta x = -ax$$

we are really stating that

$$\frac{x(n+1) - x(n)}{\tau} = -ax(n)$$

We can solve this expression to get a direct value of $x(n + 1)$ from $x(n)$:

$$x(n + 1) = (1 - a\tau)x(n)$$

Knowing the initial value of x, $x(0)$, we can deduce

$$x(1) = (1 - a\tau)x(0)$$

$$x(2) = (1 - a\tau)^2 x(0)$$

and so on until

$$x(n) = (1 - a\tau)^n x(0)$$

Now if $(1 - a\tau)$ has a magnitude less than unity, then the magnitude of x will dwindle away with time. If it is greater than unity, however, the behaviour of x will be unstable. Here again is the condition that a must be less than $2/\tau$.

Perhaps instead of a difference operator, we should be looking at a more straightforward operator with the meaning 'next'. Then 'next' $x(n)$ will be $x(n + 1)$. The example above would appear as

$$\text{'next' } x(n) = (1 - a\tau)x(n)$$

and we could even leave out the time count and just state that

$$\text{'next' } x = (1 - a\tau)x$$

This could possibly simplify first-order problems, but any system with a second-order or higher derivative will relate to a difference equation involving $x(n + 2)$ or beyond. What do we do then? In the continuous case, we were able to break a high-order differential equation down into first-order equations. Can we do the same with difference equations? Yes, if we are prepared to define some state variables.

Suppose we have a second-order difference equation

$$y(n + 2) + 4y(n + 1) + 3y(n) = 0 \tag{P10-1}$$

so that

$$y(n + 2) = -3y(n) - 4y(n + 1)$$

Let us define

$$x_1 = y(n)$$

$$x_2 = y(n + 1)$$

Then we can write two first-order difference equations

$$\text{'next' } x_1 = x_2$$

$$\text{'next' } x_2 = -3x_1 - 4x_2$$

Before we know it, we have introduced a matrix so that

$$\text{'next'} \begin{bmatrix} x_1 \\ x_2 \end{bmatrix} = \begin{bmatrix} 0 & 1 \\ -3 & -4 \end{bmatrix} \begin{bmatrix} x_1 \\ x_2 \end{bmatrix}$$

and we are back in the world of eigenvectors and eigenvalues. If we can find an eigenvector x with its eigenvalue λ, then it will have the property that

$$\text{'next'}\ x = \lambda x$$

representing stable decay or unstable growth according to the magnitude of λ. Note carefully that it is now the magnitude of λ, not its real part, that defines stability. The stable region is not the left half-plane, as before, but is the interior of a circle of unit radius centred on the origin. Only if all the eigenvalues of the system lie within this circle will it be stable. We might just stretch a point and allow the value $(1 + j.0)$ to be called stable. With this as an eigenvalue, the system could allow a variable to stand constant without decay.

Exercise 10-3-1 *Is the example of Eq. (P10-1) stable?*

The notation of the operator 'next' is not really satisfactory. We would also like to be able to represent a complete sequence by a single symbol or expression. What can we do?

In the prelude to Chapter 6, in Sec. P6-4, we saw in passing that any chosen term in a series

$$f(z) = \sum a_r z^{-r}$$

could be picked out by multiplying by z^{n-1} and performing a contour integration. If we multiply the terms of our sequence $y(n)$ by z^{-n} and sum it, we will obtain a function of z, the z *transform* of y:

$$Y(z) = \sum_{n=0}^{\infty} y(n)z^{-n}$$

It will also have a well-defined inverse, so that we can in theory retrieve the value of $y(n)$ by performing an integration around a contour which encloses the origin. We have

$$y(n) = \oint Y(z)z^{n-1}\, dz$$

We might write the relationship between $y(n)$ and $Y(z)$ as

$$Y(z) = \mathcal{Z}y(z)$$

$$y(n) = \mathcal{Z}^{-1}Y(z)$$

What do we get if we take the inverse of $Y(z)$ multiplied by an extra z?

$$Z^{-1}\{zY(z)\} = \oint z\,Y(z)z^{n-1}\,\mathrm{d}z$$

$$= \oint Y(z)z^{(n+1)-1}\,\mathrm{d}z$$

$$= y(n+1)$$

Thus multiplying the transform by z corresponds to the operation 'next', in just the same way that s related to differentiation. Since we have transforms and variables rather than operators, we can safely perform algebra on the various expressions in any way we need. If we extend the difference equation of (P10-1) by adding an input function u, its transform becomes

$$z^2 Y(z) + 4z\,Y(z) + 3Y(z) = U(z)$$

that is

$$(z^2 + 4z + 3)Y(z) = U(z)$$

$$Y(z) = \frac{1}{z^2 + 4z + 3}\,U(z)$$

and we can look for poles of the transfer function in z just as we looked for poles in s.

If we find such a pole, p, it indicates that an output sequence proportional to p^n can be obtained for no input. Again it should be stressed that p^n grows or decays according to the magnitude of p, not the sign of its real part. For stability, all poles of the z transfer function must lie inside the unit circle.

Will it not be cumbersome to have to deal with infinite sums of terms? Not really, since most of the functions we are likely to encounter sum to give a neat analytic expression. Consider, for instance, the impulse response of a simple lag. This is a decaying exponential, $\exp(-at)$. If we sample it at intervals of τ, we will have a sequence of values

$$y = 1, e^{-a\tau}, e^{-2a\tau}, \ldots, e^{-na\tau}, \ldots$$

Taking the z transform, we have

$$Y(z) = 1 + \frac{e^{-a\tau}}{z} + \frac{e^{-2a\tau}}{z^2} + \cdots$$

$$= \sum_{n=0}^{\infty} e^{-na\tau} z^{-n}$$

$$= \sum_{n=0}^{\infty} (e^{-a\tau} z^{-1})^n$$

This is now the sum of a geometric series, so we have

$$Y(z) = \frac{1}{1 - e^{-a\tau} z^{-1}}$$

We can take the z transform of a function of time, t, by summing the appropriate series in $1/z$. In the example above, we can say that

$$\mathcal{Z}(e^{-at}) = \frac{1}{1 - e^{-a\tau} z^{-1}}$$

Be careful! Although this appears to be the transform relationship between a function of time and a function of z, all the transform is concerned with is the samples at intervals of τ. When the inverse is taken, nothing at all can be said about the function at intervening values of time.

Nevertheless, we can build up a useful table of transforms, relating time functions to both Laplace and z transforms alike.

Exercise 10-3-2 What is the z transform of the impulse response of the continuous system whose transfer function is $1/(s^2 + 3s + 2)$? (First solve for the time response.) Is it the product of the z transforms corresponding to $1/(s + 1)$ and $1/(s + 2)$?

P10-4 INITIAL AND FINAL VALUE THEOREMS

With the Laplace transform, we found it useful to predict some of the properties of the time function without having to invert $F(s)$. In particular, we could calculate the value of the function just after $t = 0$ by looking at the limit of $sF(s)$ as s tended to infinity. We could also calculate the value to which the function tended after infinite time by letting s tend to zero in $sF(s)$. Can we find similar tools for the z transform?

The initial value theorem could not really be easier! If we let z tend to infinity, then in the summation every term but the first will be multiplied by a factor that tends to zero:

$$f(0) = \lim_{z \to \infty} F(z) \qquad \text{(P10-2)}$$

The final value theorem is a little more difficult. First we must stipulate that $F(z)$ has no poles on or outside the unit circle; otherwise it will not represent a function that settles to a steady value.

As z tends to unity, the function $G(z)$ will tend to the sum of all the elements of $g(n)$. If we can find a series, therefore, whose sum to n is the sequence $f(n)$, then the sum to infinity of $g(n)$ will give us the limiting value of

$f(n)$. In algebraic terms, if

$$g(n) = f(n) - f(n-1)$$

then

$$f(n) = f(-1) + \sum_0^n g(n)$$

But $f(-1)$ is zero, so

$$\underset{n \to \infty}{\text{limit}} f(n) = \sum_0^\infty g(n)$$

$$= \underset{z \to 1}{\text{limit}} G(z)$$

However,

$$g(n) = f(n) - f(n-1)$$

so

$$G(z) = F(z) - z^{-1}F(z)$$

$$= (1 - z^{-1})F(z)$$

and the final value theorem becomes

$$\underset{n \to \infty}{\text{limit}} f(n) = \underset{z \to 1}{\text{limit}} (1 - z^{-1})F(z) \qquad \text{(P10-3)}$$

P10-5 CONCLUSION

We have found a transform method to deal with sequences of sampled values, which we hope will shed light on the problems of computer control. The transform appears to be closely related to the Laplace transform in some ways, but has some important differences and pitfalls.

The earlier beta operator arose from a rough and ready discretization of differentiation, with an inverse equivalent to Euler integration. Later we will meet a better approximation which will allow us to derive computer strategies corresponding closely to a choice of s plane poles. These are both approximations, however, and rely on a sequence of computer samples being 'almost as good as the real thing'.

The z transform represents no approximation whatsoever. It describes a sequence of values and allows relations to other sequences of values to be expressed precisely, assuming only that the sampling takes place on a common heartbeat. We can build it into discrete transfer functions, enabling us to analyse stability or to design filters and compensators. Only when sampled and continuous systems are mixed do we have to take especial care.

TEN

DISCRETE TIME AND COMPUTER CONTROL

10-1 INTRODUCTION

When dedicated to a real-time control task, a computer measures a number of system variables, computes a control action and applies a corrective input to the system. It does this not continuously but at discrete instants of time. Some processes need frequent correction, such as the attitude of an aircraft, whereas the pumps and levels of a sewage process might only need attention every five minutes. Provided the corrective action is sufficiently frequent, there seems on the surface to be no reason for insisting that the intervals be regular.

In the prelude to this chapter we met the z transform, a method of analysing samples taken at regular intervals of time. The length of the sampling interval is also involved as a dividing factor in the operation of differentiation. We really need no more excuse for insisting on a regular measurement and correction cycle than that it would be more difficult to design and analyse a system that operated otherwise.

We have seen the foundations of control theory built up on the concepts of differentiation and integration. Yet here we are suggesting a change of direction to use the sample-time delay as our cornerstone. How easily can we carry the results we have established for continuous control across into discrete time? Already we have seen that eigenvectors and eigenvalues have their part to play, although with a different criterion for stability. Soon we will see that transfer functions, convolution, characteristic equations, root locus, pole placement, controllability and observability all have their very similar counterparts, among many other features.

Our first task is to relate the continuous state equations to the discrete-time behaviour.

10-2 STATE TRANSITION

The continuous system in question is described by the state equations

$$\dot{\mathbf{x}} = A\mathbf{x} + B\mathbf{u} \tag{10-1}$$

The input is driven by a digital-to-analogue converter, controlled by the output of a computer. A value of u is set up at time t and remains constant until the next output cycle at $t + \tau$. What value will the state reach at time $t + \tau$? In the working which follows, we will at first simplify matters by taking the initial time, t, to be zero.

We can solve the system equations by the *integrating factor* method, provided we can find a matrix $\exp(-At)$ whose derivative is $\exp(-At)(-A)$. Such a matrix was described in Sec. P9-4, and with its help we can use the method introduced in Sec. 2-4, now strengthened to handle a system of any order.

The method hinges on the result of differentiating $\exp(-At)x$:

$$\frac{d}{dt}(e^{-At}\mathbf{x}) = e^{-At}\dot{\mathbf{x}} + \frac{d}{dt}(e^{-At})\mathbf{x}$$

$$= e^{-At}\dot{\mathbf{x}} + e^{-At}(-A)\mathbf{x}$$

$$= e^{-At}(\dot{\mathbf{x}} - A\mathbf{x}) \tag{10-2}$$

The state equations, (10-1), tell us that

$$\dot{\mathbf{x}} - A\mathbf{x} = B\mathbf{u}$$

so

$$e^{-At}(\dot{\mathbf{x}} - A\mathbf{x}) = e^{-At}B\mathbf{u}$$

Thus with the help of (10-2) we see that

$$\frac{d}{dt}(e^{-At}\mathbf{x}) = e^{-At}B\mathbf{u} \tag{10-3}$$

Now we can integrate (10-3) to obtain

$$[e^{-At}\mathbf{x}]_0^\tau = \int_0^\tau e^{-At}B\mathbf{u}\, dt$$

that is

$$e^{-A\tau}\, \mathbf{x}(\tau) - \mathbf{x}(0) = \int_0^\tau e^{-At} B\mathbf{u}\, dt$$

The result of the product $\exp(At)\exp(-At)$ is just the unit matrix, so we can move $x(0)$ across the equation and multiply through by $\exp(At)$ to reveal that

$$\mathbf{x}(\tau) = e^{A\tau}\, \mathbf{x}(0) + e^{A\tau} \int_0^\tau e^{-At} B\mathbf{u}\, dt \qquad (10\text{-}4)$$

First notice that if the input is zero, each 'next' x is obtained simply by multiplying the present state by a matrix, $\exp(A\tau)$. This matrix is the *state transition matrix*.

If $\mathbf{u}$ is non-zero, it appears that we could have a laborious integration to perform each time. But remember that we have stipulated that $\mathbf{u}$ is set by the computer and is constant between sample times. We can therefore take $\mathbf{u}$ outside the integration. The integral can be evaluated once and for all to give a matrix which is constant for any given value of τ. We see that

$$\mathbf{x}(\tau) = M\mathbf{x}(0) + N\mathbf{u}(0) \qquad (10\text{-}5)$$

where M is the state-transition matrix and

$$N = e^{A\tau} \int_0^\tau e^{-At} B\, dt$$

This is exactly the sort of expression we need for analysing sequences of samples. We write $\mathbf{x}(n)$ to represent $\mathbf{x}$ at time $n\tau$, since an interval τ elapses between one sample and the next. Equation (10-5) then becomes

$$\mathbf{x}(n+1) = M\mathbf{x}(n) + N\mathbf{u}(n)$$

while the system output is still given by

$$\mathbf{y}(n) = C\mathbf{x}(n)$$

Our only practical problem is how to calculate M and N from the continuous state equation matrices A and B for any given τ.

10-3 DISCRETE STATE EQUATIONS AND FEEDBACK

As long as there is a risk of confusion between the matrices of the discrete state equations and those of the continuous ones, we will use the notation M and N. (Some authors use A and B in both cases, although the matrices have different values.)

If our computer is to provide some feedback control action, this must be based on measuring the system output $\mathbf{y}(n)$, taking into account a command input $\mathbf{v}(n)$ and computing an input value $\mathbf{u}(n)$ with which to drive the digital-to-analogue converters. For now we will assume that the computation is performed instantaneously as far as the system is concerned, i.e. the intervals are much longer than the computing time. We see that if the action is linear,

$$\mathbf{u}(n) = F\mathbf{y}(n) + G\mathbf{v}(n)$$

As in the continuous case, we can substitute the expression for $\mathbf{u}$ back into the system equations to get

$$\begin{aligned}
\mathbf{x}(n + 1) &= M\mathbf{x}(n) + N[F\mathbf{y}(n) + G\mathbf{v}(n)] \\
&= M\mathbf{x}(n) + N[FC\mathbf{x}(n) + G\mathbf{v}(n)] \\
&= (M + NFC)\mathbf{x}(n) + G\mathbf{v}(n)
\end{aligned}$$

Exactly as in the continuous case, we see the system matrix modified by feedback to describe a different performance. Just as before, we wish to know how to ensure that the feedback changes the performance to represent a 'better' system.

Can we still add dynamics to the controller, now operating in discrete time? Certainly; that is one of the important aims of this entire topic. Again as in the continuous case, the controller can have states of its own. It would be wise to avoid using $\mathbf{z}$ as a symbol for this state vector, because of the risk of confusion with the z transform, so we write

$$\mathbf{w}(n + 1) = K\mathbf{w}(n) + L\mathbf{y}(n) + P\mathbf{v}(n)$$

for the controller equations. The input which we apply to the system is a mixture of feedback, the controller state and the command input given by

$$\mathbf{u}(n) = F\mathbf{y}(n) + G\mathbf{w}(n) + H\mathbf{v}(n)$$

When we combine these with the system equations

$$\mathbf{x}(n + 1) = M\mathbf{x}(n) + N\mathbf{u}(n)$$

$$\mathbf{y}(n) = C\mathbf{x}(n)$$

we arrive at

$$\begin{bmatrix} \mathbf{x}(n + 1) \\ \mathbf{w}(n + 1) \end{bmatrix} = \begin{bmatrix} M + NFC & NG \\ \hline LC & K \end{bmatrix} \begin{bmatrix} \mathbf{x}(n) \\ \mathbf{w}(n) \end{bmatrix} + \begin{bmatrix} NH \\ P \end{bmatrix} \mathbf{v}(n)$$

As before, this composite matrix state equation can cover all the cases of simple feedback, feedback with dynamics and dynamics applied to the command signal. Provided the system is controllable and observable, the poles can in principle be placed anywhere.

Just how do we examine for controllability in discrete terms? The tests are much the same, but the arguments for their validity are easier to justify than for the continuous case. Let us first look at controllability. If we start from an initial zero state, then after the first interval

$$\mathbf{x}(1) = N\mathbf{u}(0)$$

At the next sample time, having applied $u(1)$,

$$\mathbf{x}(2) = MN\mathbf{u}(0) + N\mathbf{u}(1)$$

then

$$\mathbf{x}(3) = M^2N\mathbf{u}(0) + MN\mathbf{u}(1) + N\mathbf{u}(3)$$

and so on. If we are to be able to drive $\mathbf{x}$ in any direction within its space of n dimensions, the matrix

$$(N \,|\, MN \,|\, M^2N \,|\, \cdots)$$

must have a rank equal to the number of state components. Again we can stop our search at the $M^{n-1}N$ term.

A consequence of controllability is that it is theoretically possible to drive an nth-order system to any chosen state in n sampling intervals or less.

For observability, we must first look at the task of deducing a past value of the state. We can assume that we have not only the present values of the output vector components, $\mathbf{y}$, but as many past values as we may need, together with a record of the corresponding inputs. Now

$$C\mathbf{x}(n-2) = \mathbf{y}(n-2)$$

gives us one set of equations for $x(n-2)$. The state equations tell us that

$$\mathbf{x}(n-1) = M\mathbf{x}(n-2) + N\mathbf{u}(n-2)$$

so, multiplying by C, and replacing $C\mathbf{x}(n-1)$ by $\mathbf{y}(n-1)$:

$$CM\mathbf{x}(n-2) = \mathbf{y}(n-1) - CN\mathbf{u}(n-2)$$

giving us some more equations. Then

$$\mathbf{x}(n) = M^2\mathbf{x}(n-2) + MN\mathbf{u}(n-2) + N\mathbf{u}(n-1)$$

so

$$CM^2\mathbf{x}(n-2) = \mathbf{y}(n) - CMN\mathbf{u}(n-2) - CN\mathbf{u}(n-1)$$

Putting these sets of equations together, we see that they can be solved for $\mathbf{x}(n-2)$ if the rank of the matrix

$$\begin{bmatrix} C \\ CM \\ CM^2 \end{bmatrix}$$

is equal to the order of the system. If the system is third order or less and the rank is not sufficient, then the system is not controllable. Otherwise we must

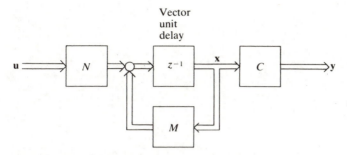

Figure 10-1 Representation of a discrete system.

try instead to deduce an earlier state and add another block or more to the matrix in question.

Having succeeded in deducing a past value of the state, we can substitute that value and the corresponding value of input into the state equations to calculate the next value, and so on until we have reached the present. Just as the observer in the continuous case needed integrators to operate, so in the discrete case we must add memory elements to store past values of output and input.

In this section, we have expressed the system dynamics in terms of one-sample delays. We can represent a simulation of the system in block diagram form, just as we did in the continuous case, but the 'boxes' here will be unit delays instead of integrators. For want of a better symbol we will use z^{-1} to label these delays, resulting in the diagram of Fig. 10-1.

We were able to simulate the continuous state equations on a digital computer, achieving a close approximation to the system behaviour. Now we can simulate the discrete state equations exactly. There is no approximation; indeed, part of the system may itself represent a computer program. In the code which follows, remember that the index in brackets is not the sample number, but the index of the particular component of the state or input vector. The sample time is 'now'.

Assume that all variables have been declared and initialized, and that we are concerned with computing the next value of the state knowing the input **u**. The state has n components and there are m components of input. For the coefficients of the discrete state matrix we will use $A(I, J)$, since there is no risk of confusion, and we will use $B(I, J)$ for the input matrix:

```
FOR I = 1 TO N

  NEWX(I) = 0

  FOR J = 1 TO N
    NEWX(I) = NEWX(I) + A(I, J)*X(J)
  NEXT J
```

```
FOR J = 1 TO M
   NEWX(I) = NEWX(I) + B(I,J)*U(J)
NEXT J

NEXT I

FOR I = 1 TO N
  X(I) = NEWX(I)
NEXT I
```

Written in matrix form, all the problems appear incredibly easy. That is how it should be. A little more effort is needed to set up discrete equations corresponding to a continuous system, but even that is quite straightforward. Let us consider an exercise. Try it first, then read the next section to see it worked through.

Exercise 10-3-1 *The output position of a servomotor is related to the input by a continuous transfer function $a/[s(s + 1)]$, i.e. when driven from one volt the output accelerates to a speed of a radians per second, with a settling time of one second. Express this in state-variable terms. It is proposed to apply computer control by sampling the error at intervals of 0.1 second and applying a proportional corrective drive to the motor. Choose a suitable gain and discuss the response to an initial error. The system is illustrated in Fig. 10-2.*

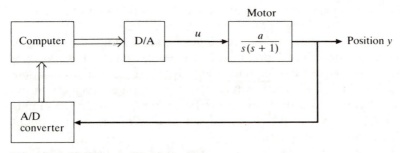

Figure 10-2 Computer position control.

10-4 A PRACTICAL EXAMPLE OF DISCRETE-TIME CONTROL

Let us work through Exercise 10-3-1. The most obvious state variables of the motor are position and velocity. The proposed feedback control uses position alone; we will assume that the velocity is not available as an output. We have

$$\dot{x}_1 = x_2$$

and
$$\dot{x}_2 = -x_2 + au$$

while
$$y = x_1$$

We could attack the time solution by finding eigenvectors and diagonalizing the matrix, then taking the exponential of the matrix and transforming it back, but that is really overkill in this case. We can apply 'knife-and-fork' methods to see that if u is constant, x_2 is of the form $b \exp(-t) + c$. Substituting into the differential equation and matching initial conditions gives

$$x_2(t + \tau) = x_2(t)\,e^{-\tau} + a(1 - e^{-\tau})u$$

$x_1(t + \tau)$ is found from the integral of x_2 to be

$$x_1(t + \tau) = x_1(t) + x_2(t)(1 - e^{-\tau}) + a(\tau - 1 + e^{-\tau})u$$

If we take τ to be 0.1 seconds, we can put numbers to the coefficients. As $\exp(-0.1) = 0.905$, we have

$$\begin{bmatrix} x_1(n + 1) \\ x_2(n + 1) \end{bmatrix} = \begin{bmatrix} 1 & 0.095 \\ 0 & 0.905 \end{bmatrix} \begin{bmatrix} x_1(n) \\ x_2(n) \end{bmatrix} + \begin{bmatrix} 0.005a \\ 0.095a \end{bmatrix} u(n)$$

Now we propose to apply discrete feedback around the loop, making $u(n) = -kx_1(n)$, and we find that

$$\begin{bmatrix} x_1(n + 1) \\ x_2(n + 1) \end{bmatrix} = \left\{ \begin{bmatrix} 1 & 0.095 \\ 0 & 0.905 \end{bmatrix} + \begin{bmatrix} 0.005a \\ 0.095a \end{bmatrix} \begin{bmatrix} -k & 0 \end{bmatrix} \right\} \begin{bmatrix} x_1(n) \\ x_2(n) \end{bmatrix}$$

that is

$$\begin{bmatrix} x_1(n + 1) \\ x_2(n + 1) \end{bmatrix} = \begin{bmatrix} 1 - 0.005ak & 0.095 \\ -0.095ak & 0.905 \end{bmatrix} \begin{bmatrix} x_1(n) \\ x_2(n) \end{bmatrix}$$

Next we must examine the eigenvalues, and choose a suitable value for k:

$$\det \begin{vmatrix} 1 - 0.005ak - \lambda & 0.095 \\ -0.095ak & 0.905 - \lambda \end{vmatrix} = 0$$

so we have

$$\lambda^2 + (0.005ak - 1.905)\lambda + (0.905 + 0.0045ak) = 0$$

If $k = 0$, the roots are 1 and 0.905. If ak exceeds 20 we are in danger of making the constant term exceed unity; this is equal to the product of the roots, so must be less than one for stability. Just where within this range should we place k? Suppose we look for equal roots; then we have

$$(0.005ak - 1.905)^2 = 4(0.905 + 0.0045ak)$$

The smaller solution is the acceptable one, and here

$$0.005ak = 0.001215$$

giving ak a value of 0.243—very much less than 20! The roots will now be at 0.9519, indicating that after two seconds the disturbance will have decayed to $(0.9519)^{20} = 0.373$ of its original value.

This is not a surprising result. It matches closely the response obtained for continuous feedback when the same system was examined in Sec. 7-3.

Try working through the next exercise before reading on to see its solution.

Exercise 10-4-1 What happens if we sample and correct the system of Exercise 10-3-1 less frequently, say at one second intervals? Find the new discrete matrix state equations, the feedback gain for equal eigenvalues and the response after two seconds.

The discrete equations were found to be

$$\mathbf{x}(n + 1) = \begin{bmatrix} 1 & 1 - e^{-\tau} \\ 0 & e^{-\tau} \end{bmatrix} \mathbf{x}(n) + \begin{bmatrix} a(\tau + e^{-\tau} - 1) \\ a(1 - e^{-\tau}) \end{bmatrix} u(n) \quad (10\text{-}6)$$

With τ set to unity, these become

$$\mathbf{x}(n + 1) = \begin{bmatrix} 1 & 0.63212 \\ 0 & 0.36788 \end{bmatrix} \mathbf{x}(n) + \begin{bmatrix} 0.36788a \\ 0.63212a \end{bmatrix} u(n)$$

With feedback gain k we have

$$\mathbf{x}(n + 1) = \begin{bmatrix} 1 - 0.36788ak & 0.63212 \\ -0.63212ak & 0.36788 \end{bmatrix} \mathbf{x}(n)$$

yielding a characteristic equation

$$\lambda^2 + (0.36788ak - 1.36788)\lambda + (0.26424ak + 0.36788) = 0$$

The limit of ak for stability has now reduced below 2.4 (otherwise the product of the roots is greater than unity), and for equal roots we have

$$(0.36788ak - 1.36788)^2 = 4(0.26424ak + 0.36788)$$

Pounding a pocket calculator gives $ak = 0.196174$—smaller than ever! With this value substituted, the eigenvalues both become 0.647856. It seems an improvement on the previous eigenvalue, but remember that this time it applies to one second of settling. In two seconds, the error is reduced to 0.4197 of its initial value, only a little worse than the previous control when correction was made ten times per second.

Can we really get away with correcting this system as infrequently as once per second? If the system performs exactly as its equations predict, then it appears possible. In practice, however, there is always uncertainty in any system. A position servo is concerned with achieving the command position, but it must also maintain that position in the presence of unknown disturbing forces which can arrive at any time. A once-per-second correction to the flight control surfaces of an aircraft might suffice for straight and level flight in calm air, but in turbulence this could be disastrously inadequate.

10-5 DEAD-BEAT RESPONSE

In the linear case, when we allowed all the states to appear as outputs for feedback purposes, we found that we could put the poles of the closed loop system wherever we wished. What is the corresponding consequence in the discrete case? We find that in most cases we can achieve settling in at most a number of samples corresponding to the order of the system. This is not settling as we have met it in continuous time, with exponential decays which never quite reach zero. This is settling in the grand style, with all states reaching an exact zero and remaining there from that sample onwards. The response is dead-beat.

Take the example of Exercise 10-3-1, now with samples being taken of motor velocity as well as position. When we examined the problem numerically we unearthed a torrent of digits. Let us generalize the coefficients and say that the state equation (10-6) amounts to

$$\mathbf{x}(n + 1) = \begin{bmatrix} 1 & p \\ 0 & q \end{bmatrix} \mathbf{x}(n) + \begin{bmatrix} r \\ s \end{bmatrix} u(n)$$

(Here s is just a constant—nothing to do with Laplace!) Now we have both states to feed back, so we can make

$$u(n) = (a \quad b)\mathbf{x}(n)$$

leading to a closed loop state equation

$$\mathbf{x}(n + 1) = \begin{bmatrix} 1 + ar & p + br \\ as & p + bs \end{bmatrix} \mathbf{x}(n)$$

The eigenvalues are given by

$$\lambda^2 - (1 + ar + p + bs)\lambda + (p + bp + arp - asp) = 0$$

What is to stop us choosing a and b such that both these coefficients are zero? Nothing. We can assign values that make both eigenvalues zero by making

$$(1 + p) + ar + bs = 0$$

and

$$p + ap(r - s) + bs = 0$$

The closed loop matrix that results is of the form

$$\begin{bmatrix} c & d \\ -c^2/d & -c \end{bmatrix}$$

Its rank is only one, and when multiplied by itself gives the zero matrix—try filling in the details.

In this particular case, the state will settle neatly to zero in the continuous sense too. That need not always be the case, particularly when the system has a damped oscillatory tendency at a frequency near a multiple of the sampling frequency. All that is assured is that the values read at the sampling instants will be zero, and care should always be taken to check the continuous equations for the possibility of trouble in between.

Notice that the designer is faced with an interesting choice. A dead-beat system has a transition matrix with rank less than the order of the system. The square of this matrix has an even lower rank, and at each multiplication the rank must reduce by at least one if it is ultimately to reach zero. Thus a third-order dead-beat system must settle in at most three sampling intervals. Within the limitations of drive amplifiers, a certain settling time is reasonable for a typical disturbance and any faster performance will be dominated by amplifier saturation.

The designer must decide whether to shoot for a dead-beat response, which could therefore require a low sampling rate, or whether to sample rapidly and be satisfied with an exponential style of response. Of course saturation need be no bad thing, but the design must then take it into account. The presence or absence of noise will help in making the decision. Also, to achieve a dead-beat performance the designer must have an accurate knowledge of the open loop parameters. Frequent correction is usually safest unless the system contains transport delays.

Returning to our example, we have seen the use of observers in the continuous case for constructing 'missing' feedback variables. What is the equivalent in discrete time?

10-6 DISCRETE-TIME OBSERVERS

In continuous time we made observers with the aid of integrators; now we will try to achieve a similar result with time delays—perhaps with a few lines of software. First consider the equivalent of the Kalman filter, in which the observer contains a complete model of the system. The physical system is

$$\mathbf{x}(n + 1) = M\mathbf{x}(n) + N\mathbf{u}(n)$$

$$\mathbf{y}(n) = C\mathbf{x}(n)$$

while the model is

$$\hat{\mathbf{x}}(n + 1) = M\hat{\mathbf{x}}(n) + N\mathbf{u}(n) + K[\mathbf{y}(n) - \hat{\mathbf{y}}(n)]$$

$$\hat{\mathbf{y}}(n) = C\hat{\mathbf{x}}(n)$$

Just as in the continuous case, there is a matrix equation to express the discrepancy between the estimated states; here it is a difference equation ruled by the matrix $M + KC$. If the system is observable we can choose coefficients of K to place the eigenvalues of $M + KC$ wherever we wish. In particular, we should be able to make the estimation dead-beat, so that after two or three intervals an error-free estimate is obtained.

We would also like to look closely at reduced-state observers, perhaps to find the equivalent of phase advance. Now we set up a system within the controller which has state $\mathbf{w}$, where

$$\mathbf{w}(n + 1) = P\mathbf{w}(n) + Q\mathbf{u}(n) + R\mathbf{y}(n)$$

Now we would like $\mathbf{w}(n)$ to approach some mixture of the system states, $S\mathbf{x}(n)$. We look at the equations describing the future difference between these values, obtaining

$$\mathbf{w}(n + 1) - S\mathbf{x}(n + 1) = P\mathbf{w}(n) + Q\mathbf{u}(n) + R\mathbf{y}(n) - SM\mathbf{x}(n) - SN\mathbf{u}(n)$$

$$= P\mathbf{w}(n) + (RC - SM)\mathbf{x}(n) + (Q - SN)\mathbf{u}(n) \quad (10\text{-}7)$$

[using $\mathbf{y}(n) = C\mathbf{x}(n)$]. If we can bend the equations into the form

$$(\mathbf{w} - S\mathbf{x})_{(n + 1)} = P(\mathbf{w} - S\mathbf{x})_{(n)}$$

where P describes a system whose signals decay to zero, then we will have succeeded. To achieve this, the right-hand side of Eq. (10-7) must reduce to

$$P[\mathbf{w}(n) - S\mathbf{x}(n)]$$

requiring

$$RC - SM = PS$$

and

$$Q - SN = 0 \quad (10\text{-}8)$$

Let us tie together observers and response design in one simple example. By modifying the motor position control example so that the motor is undamped, we make the engineering task more challenging, at the same time making the arithmetic very much simpler.

Exercise 10-6-1 *The motor of a position control system satisfies the continuous differential equation $\ddot{y} = u$. The loop is to be closed by a computer which samples the position and outputs a drive signal at regular intervals. Design an appropriate control algorithm to achieve settling within one second.*

First we work out the discrete-time equations. This can be done from first principles, without calling on matrix techniques. The state variables are position and velocity. The change in velocity between samples is $u\tau$, so

$$x_2(n + 1) = x_2(n) + \tau u(n)$$

Integrating the velocity equation gives

$$x_1(n + 1) = x_1(n) + \tau x_2(n) + \frac{\tau^2}{2} u(n)$$

so we have

$$\mathbf{x}(n + 1) = \begin{bmatrix} 1 & \tau \\ 0 & 1 \end{bmatrix} \mathbf{x}(n) \begin{bmatrix} \tau^2/2 \\ \tau \end{bmatrix} u(n) \tag{10-9}$$

Let us first try simple feedback, making $u(n) = ky(n)$. Then the closed loop response to a disturbance is governed by

$$\mathbf{x}(n + 1) = \begin{bmatrix} 1 + k\tau^2/2 & \tau \\ k\tau & 1 \end{bmatrix} \mathbf{x}(n)$$

The eigenvalues are the roots of

$$\det \begin{vmatrix} 1 + k\tau^2/2 - \lambda & \tau \\ k\tau & 1 - \lambda \end{vmatrix} = 0$$

that is

$$\lambda^2 - \left(2 + \frac{k\tau^2}{2} \right) \lambda + 1 - \frac{k\tau^2}{2} = 0$$

If k is positive, the sum of the roots (the negative of the coefficient of λ) is greater than two; if k is negative, the product of the roots (the constant term) is greater than one; we cannot win. The controller will therefore have to contain some dynamics.

If we had the velocity available for feedback, we could set $u = ax_1 + bx_2$. That would give us a closed loop matrix

$$\begin{bmatrix} 1 + a\tau^2/2 & \tau + b\tau^2/2 \\ a\tau & 1 + b\tau \end{bmatrix}$$

For the eigenvalues, we again subtract λ from each diagonal term and take the determinant. Now if we want to try for a dead-beat response we need both eigenvalues zero, so the coefficients of λ and of the constant term must vanish:

$$1 + \frac{a\tau^2}{2} + 1 + b\tau = 0$$

and

$$\left(1 + \frac{a\tau^2}{2} \right)(1 + b\tau) - a\tau \left(\tau + \frac{b\tau^2}{2} \right) = 0$$

that is

$$2 + \frac{a\tau^2}{2} + b\tau = 0$$

and

$$1 - \frac{a\tau^2}{2} + b\tau = 0$$

so

$$a\tau^2 = -1$$

and

$$b\tau = -1.5$$

Thus

$$u(n) = -\frac{x_1(n) + 1.5\tau x_2(n)}{\tau^2} \tag{10-10}$$

So where are we going to find a velocity signal? From a dead-beat observer. Set up a first-order system with one state variable $w(n)$, governed by

$$w(n + 1) = pw(n) + qy(n) + ru(n)$$

The state will estimate $x_1(n) + kx_2(n)$ if [see Eq. (10-7)]

$$rC - (1 \qquad k)M = p(1 \qquad k)$$

and

$$q - (1 \qquad k)N = 0$$

where M and N are the system matrices of Eq. (10-9).

If we are going to try for dead-beat estimation, p will be zero. We now have

$$(r \qquad 0) - (1 \qquad \tau + k) = (0 \qquad 0)$$

and

$$q - \left(\frac{\tau^2}{2} + k\tau\right) = 0$$

From the first of these, $k = -\tau$ and $r = 1$. Substituting in the second, $q = -\tau^2/2$.

Now if we make

$$w(n + 1) = -\frac{\tau^2}{2} u(n) + y(n) \tag{10-11}$$

then after one interval w we will have the value $x_1 - \tau x_2$.

For the feedback requirements of Eq. (10-10) we need

$$u(n) = -\frac{x_1(n) + 1.5\tau x_2(n)}{\tau^2}$$

so we use a mix of $1.5w - 2.5x_1$ to get

$$u(n) = \frac{3w(n) - 5y(n)}{2\tau^2} \tag{10-12}$$

Now we can reexamine this result in terms of its discrete transfer function. In the prelude to this chapter, the z transform was introduced as a convenient way of dealing with sample delays. The observer of Eq. (10-12) can be written in transform form as

$$zW = \frac{-\tau^2}{2} U + Y$$

and it can be combined with the controller

$$U = \frac{3W - 5Y}{2\tau^2}$$

to give

$$U = \frac{3[(-\tau^2/2)U + Y]/z - 5Y}{2\tau^2}$$

or

$$4\tau^2 U = -\frac{3\tau^2 U}{z} - 10Y + \frac{6Y}{z}$$

(10-13)

that is

$$U = -\frac{10 - 6/z}{(4 + 3/z)\tau^2} Y$$

(10-14)

This transfer function does in the discrete case what phase advance does for the continuous system. It is interesting to look at its *impulse* response, the result of applying an input sequence $(1, 0, 0, \ldots)$. Let us assume that $\tau = 1$. Equations (10-13) can be used as a 'recipe' for u in terms of its last value u/z, of y and of the previous y, y/z. We arrive at the sequence

$$-2.5, \ 3.375, \ -2.53125, \ 1.8984, \ -1.4238, \ 1.0679, \ldots$$

After the initial transient, each term is -0.75 times the previous one. The step response, the result of applying $(1, 1, 1, \ldots)$, is

$$-2.5, \ 0.875, \ -1.65625, \ 0.2422, \ -1.1816, \ -0.1138, \ldots$$

The impulse response sequence is illustrated in Fig. 10-3.

Having designed the controller, how do we implement it?

Let us apply control at intervals of 0.5 seconds, so that $\tau^2 = 0.25$, and let us use the observer state w in the algorithm. Within the control software, the variable W must be updated from measurements of the output Y and from the previous drive value U. First it must be used in computing the new drive:

```
U = 6*W - 10*Y
```

Then it is updated by assigning

```
W = Y - 0.125U
```

Now U must be output to the digital-to-analogue converter and a process must be primed to perform the next correction after a delay of 0.5 seconds. Where have the dynamics of the controller vanished to?

The second assignment does not really assign a value to w, but to the 'next' value of w, zw. This value is remembered until the next cycle, and it is this memory which forms the dynamic effect.

To test the effectiveness of the controller, it is not hard to simulate the entire system—in this very simple case we will even avoid using arrays. Let us

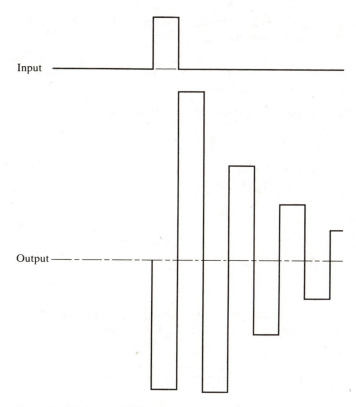

Input

Output

Figure 10-3 Response of discrete-time 'phase advance'.

start with all states zero, including the controller, and then apply a disturbance to the motor position. First we evaluate some coefficients of the state transition matrix, such as it is:

```
DT  = 0.5
A11 = 1
A12 = DT
A22 = 1
B11 = DT*DT/2
B12 = DT
```

Now we disturb the initial position

```
X1 = 1
```

and enter the simulation loop, stepping the simulation time:

```
FOR T = 0 TO 5 STEP DT
Y  = X1
U  = (3*W − 5*Y)/(2*DT*DT)
W = Y − U*DT*DT/2
```

That deals with the controller. Before updating the motor, let us print or plot its deflection with some command such as

PRINT T, X1

or

LINE − (ORIGINT + TSCALE∗T, ORIGINX + XSCALE∗X1)

Now for the motor simulation:

NEWX1 = A11∗X1 + A12∗X2 + B11∗U
NEWX2 =   ASS∗X2 + B12∗U
X1   = NEWX1
X2   = NEWX2

(The introduction of NEWX1 and NEWX2 is not really necessary in this particular case.) We complete the program with

NEXT T

Exercise 10-6-2   *Tailor the above program to your particular computer and run the simulation for a variety of values of DT. Now alter the motor parameter*

```
5  TØ=4Ø:TSC=1ØØ:XØ=1ØØ:XSC=-8Ø:SCREEN 2
1Ø DT=.5
2Ø A11=1:A12=DT:A22=1
25 CLS
3Ø INPUT"mismatch factor ",K
4Ø B11=K*DT*DT/2
5Ø B12=K*DT
6Ø X1=1
62 LINE(TØ+5*TSC,XØ)-(TØ,XØ)
65 LINE(TØ,XØ-XSC)-(TØ,XØ+X1*XSC)
7Ø FOR T=Ø TO 5 STEP DT
1ØØ Y=X1
11Ø U=(3*W-5*Y)/(2*DT*DT)
12Ø W=Y-U*DT*DT/2
2ØØ REM PRINT T,X1
22Ø LINE - (TØ+TSC*T, XØ+XSC*X1)
3ØØ X1=A11*X1+A12*X2+B11*U
31Ø X2=        A22*X2+B12*U
32Ø NEXT T
```

Figure 10-4 Software which produced Fig. 10-5.

from the assumed value (one unit per second per second) by changing the values of B11 and B12 by a factor K: replace the lines which define them by

```
INPUT "MISMATCH K "; K
B11 = K*DT*DT/2
B12 = K*DT
```

How much variation of K from unity is possible before the performance becomes unacceptable?

The results of running the program listed in Fig. 10-4 are shown in Fig. 10-5. When the observer is well matched to the system, it is clear that a disturbance can be corrected completely and perfectly in just three sampling intervals, i.e. 1.5 seconds. A motor with a lower acceleration than predicted will take disproportionately much longer to settle, with damping which deteriorates as the acceleration is reduced.

If the motor is unexpectedly agile, however, the result can be disastrous.

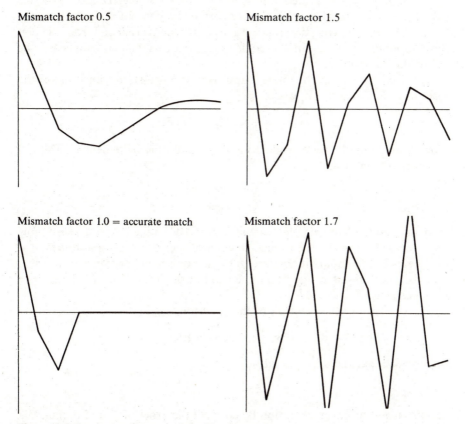

Mismatch factor 0.5

Mismatch factor 1.5

Mismatch factor 1.0 = accurate match

Mismatch factor 1.7

Figure 10-5 Responses of attempted dead-beat control.

Any slight underestimate of acceleration will give a damped oscillatory response, but when the estimate is reduced to 0.6 of the actual value (i.e. 1/1.7) then the oscillation diverges. An error of such a factor can easily occur in practice—use caution when striving for high performance!

10-7 CALCULATING THE STATE TRANSITION MATRIX

In the case of distinct eigenvalues, the task looks algebraically easy. As outlined in the prelude to Chapter 9, a transformation T allows us to form a diagonal matrix:

$$TAT^{-1} = \Lambda$$

The function $\exp(A\tau)$ can then be calculated from the matrix $\exp(\Lambda\tau)$, which is a diagonal matrix with elements $e^{\lambda_n \tau}$, by computing $T^{-1}\exp(\Lambda\tau)T$. Nevertheless, the process is far from straightforward. First the characteristic equation has to be formed and solved. Then the eigenvalues must be substituted back to allow the eigenvectors to be computed, leading to the inverse of the transformation matrix. This must be inverted and from then on the computation is straightforward. Is there a process that can more easily be left to the computer?

In Sec. 2-3 we met the crudest approximation to state transition, in the form of Euler integration, where

$$x(t + \delta t) \simeq x(t) + \dot{x}(t)\,\delta t$$

Now if, for example, we considered a very small time increment $\tau/1024$ and computed the matrix

$$I + \frac{A\tau}{1024}$$

then the result of raising this to the power 1024 might give a reasonable approximation to $\exp(A\tau)$. In fact, to achieve such a power of a matrix, it is only necessary to square it ten times. On the other hand, arithmetic rounding errors in the computer might take their toll of accuracy. What is more, there is an easier way.

If we consider the solution of the equation

$$\dot{x} = x, \qquad \text{where } x(0) = 1$$

then we see that the approximation

$$x = 1 + \tau$$

deviates from the correct solution by terms of the order of τ^2 and higher. We can obtain a better approximation as follows:

By the Euler method, estimate a midpoint value $x(\tau/2)$. At this point work out the derivative. From the starting point $c(0)$, take a full step with this new slope.

$$x(0) = 1$$

$$x\left(\frac{\tau}{2}\right) = 1 + \frac{\tau}{2}$$

$$\dot{x}\left(\frac{\tau}{2}\right) = 1 + \frac{\tau}{2}$$

$$x(\tau) = x(0) + \tau\dot{x}\left(\frac{\tau}{2}\right)$$

$$= 1 + \tau + \frac{\tau^2}{2}$$

When applied to the general case, this extra term makes a dramatic improvement to the quality of the integration. It will be seen in the next chapter that it relates to *trapezoidal integration* and to an improved transform method. It can readily be calculated for, say, $\tau/128$ and then combined with the *repeated squaring* technique to arrive at an acceptable transition matrix for τ.

There are other numerical integration methods that are appropriate, including Runge–Kutta–Gill. This in effect takes the half-way gradient method a stage further. Gradients are calculated at the starting point and at three other points, to be mixed together in a ratio which when used for the final step gives terms accurate to the fourth power of τ. Its basis is as follows:

Divide the interval into three.

Calculate the gradient at the start point X0, and with this gradient take a step of length $\tau/3$ to, say, X1.

Calculate the gradient at X1, and use this to take a step of length $2\tau/3$ from X0 to X2.

Calculate the gradient at X2, and use this to take a step of length τ from X0 to X3.

Calculate the gradient at X3.

Mix the gradients in the following proportions:

$$(DX0 + 3*DX1 + 3*DX2 + DX3)/8$$

With this new gradient, take a step of length τ from X0 to arrive at the target.

Exercise 10-7-1 *Apply this algorithm to $\dot{x} = x$ to show that the result is accurate to terms in the fourth power of τ.*

10-8 CONCLUSION

We have seen the methods of state space in continuous time transferred to sampled systems, with the possibility of designing controllers which give dead-beat responses. Many more techniques are based on the transfer function approach, including root locus, and in the next chapter we will draw links between the z transform and the methods of Laplace and Fourier.

RELATIONSHIP BETWEEN z AND OTHER TRANSFORMS

P11-1 INTRODUCTION

The z transform involves multiplying an infinite sequence of samples by a function of z and sample number and summing the resulting series to infinity. The Laplace transform involves multiplying a continuous function of time by a function of s and time and integrating the result to infinity. They are obviously closely related, and in this prelude we examine ways of finding one from the other. We also look at another transform, the w transform, which is useful as a design method for approximating discrete control to an s plane specification.

P11-2 THE IMPULSE MODULATOR

In moving between continuous and discrete time in the realm of the Laplace transform, we have to introduce an artificial conceptual device which has no real physical counterpart. This is the *impulse modulator*. At each sample time, the value of the continuous input signal is sampled and an output is defined which is an impulse having an area equal to the value of the sample. Of course the trouble arises because the time function which has a Laplace transform of unity is the unit impulse at $t = 0$.

The Laplace transform is concerned with integrations over time, and any such integral involving pulses of zero width will give zero contribution unless the pulses are of infinite height. Could these pulses be made broader and finite? A *genuine* discrete controller is likely to use a *zero-order hold*, which latches the input signal and gives a constant output equal to the sampled value, not changing until the next sample time. As we will see, although this device is fundamental to computer control it introduces dynamic problems of its own.

The unit impulse and its transform were discussed in some detail in Sec. 8-7. Let us recollect a few of their important properties.

The unit impulse is denoted by $\delta(t)$; it is zero everywhere except at $t = 0$ where it is infinite and it has an integral of unity. Its Laplace transform is

$$\mathcal{L}\{\delta(t)\} = \int_{0-}^{\infty} \delta(t)\,e^{-st}\,dt$$

$$= 1$$

(The lower limit of $0-$ denotes that the integral is taken from 'just less' than zero, to include the whole impulse.)

An impulse which is delayed until $t = \tau$ can be represented by $\delta(t - \tau)$. Now its Laplace transform is

$$\mathcal{L}\{\delta(t - \tau)\} = \int_{0-}^{\infty} \delta(t - \tau)\,e^{-st}\,dt$$

$$= e^{-s\tau}$$

A sample of the signal $x(t)$ can be taken at time $t = \tau$ by the device of multiplying $x(t)$ by the time function $\delta(t - \tau)$ and integrating the product over all time; the value and its time of sampling are well represented by the product before it is integrated.

Two samples, at $t = \tau$ and at $t = 2\tau$, can be represented by the product of $x(t)$ with $\delta(t - \tau) + \delta(t - 2\tau)$, as suggested in Fig. P11-1. Extending the concept, the train of samples of $x(t)$ at $t = 0, t = \tau, t = 2\tau, \ldots, t = n\tau, \ldots$ may be represented by the product of the function $x(t)$ with an infinite train of impulses $\delta(t) + \delta(t - \tau) + \cdots + \delta(t - n\tau) + \cdots$. This is the operation of the impulse modulator.

From an input signal $x(t)$, the impulse modulator gives an output which is the infinite train of impulses

$$\sum_{n=0}^{\infty} x(n\tau)\,\delta(t - n\tau)$$

When we take the Laplace transform of the output of the impulse modulator, we have the sum of the transforms of all the individual impulses

$$\sum_{n=0}^{\infty} x(n\tau)\,e^{-sn\tau} = \sum_{n=0}^{\infty} x(n\tau)(e^{s\tau})^{-n}$$

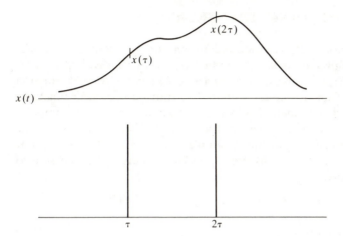

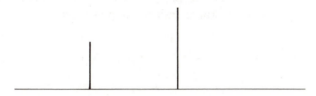

$m(t) = \delta(t - \tau) + \delta(t - 2\tau)$ — the sum of two unit impulses

$x(t)\,m(t) = x(\tau)\,\delta(t - \tau) + x(2\tau)\,\delta(t - 2\tau)$
$x(t)$ is sampled by multiplying it by an impulse train

$x(t) \longrightarrow$ | Impulse modulator | $\longrightarrow$ Sample train

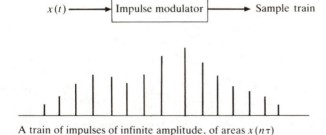

A train of impulses of infinite amplitude, of areas $x(n\tau)$

Figure P11-1 Impulse modulation.

Compare this with the z transform

$$\sum_{n=0}^{\infty} x(n\tau) z^{-n}$$

and the connection is obvious. The z transform of a function of time is found by taking the Laplace transform of the impulse-modulated function and substituting z for $\exp(s\tau)$ in each expression.

P11-3 CORRESPONDING TRANSFORMS

When one continuous filter is followed by another, the Laplace transfer function of the combination is the product of the two individual transfer functions. A lag $1/(s + a)$ following a lag $1/(s + b)$ has the transfer function $1/[(s + a)(s + b)]$. Can we similarly multiply z transform transfer functions to represent cascaded elements? When we come to look at the impulse responses, we see that the response of the first filter is e^{-at}. To derive the corresponding z transform, we must sample this time function at regular intervals, multiply the samples by corresponding powers of $1/z$ and sum to infinity. We thus have

$$(e^{-at}) = 1 + z^{-1} e^{-\tau} + z^{-2} e^{-2\tau} + \cdots + z^{-n} e^{-n\tau} + \cdots$$

$$= \frac{1}{1 - z^{-1} e^{-a\tau}} \qquad \text{(P11-1)}$$

When we derive the z transform corresponding to the combined filter, we must first invert the Laplace transform to find the impulse response.

$$\mathcal{L}^{-1}\left\{ \frac{1}{(s + a)(s + b)} \right\} = \mathcal{L}^{-1}\left\{ \frac{1}{b - a}\left(\frac{1}{s + a} - \frac{1}{s + b} \right) \right\}$$

$$= \frac{1}{b - a} (e^{-at} - e^{-bt})$$

Now sampling at intervals of τ, multiplying by inverse powers of z and summing gives us the z transform:

$$\frac{1}{b - a}\left(\frac{1}{1 - z^{-1} e^{-a\tau}} - \frac{1}{1 - z^{-1} e^{-b\tau}} \right)$$

$$= \frac{1}{b - a} \frac{z^{-1} e^{-a\tau} - z^{-1} e^{-b\tau}}{(1 - z^{-1} e^{-a\tau})(1 - z^{-1} e^{-b\tau})}$$

$$= \frac{e^{-a\tau} - e^{-b\tau}}{b - a} \frac{z^{-1}}{(1 - z^{-1} e^{-a\tau})(1 - z^{-1} e^{-b\tau})} \qquad \text{(P11-2)}$$

This is quite obviously not the product of two terms of the form of (P11-1)—there is an extra factor of

$$\frac{e^{-a\tau} - e^{-b\tau}}{b - a} z^{-1}$$

The response of the compound lag appears to have an extra sample's delay when compared with the product of the two separate z transforms. By looking at the equivalent convolution, it is not hard to see why.

When an impulse is applied to the first lag, the response is a decaying exponential in continuous time, or a sequence of decaying samples in discrete time. If a sampling device converts the sequence of values into a corresponding train of impulses and applies them to a second lag, then a second sampler will indeed find a signal as described by the product of the two z transforms.

The unit impulse applied at the input will produce a first sample of unity at the output of the first lag. This is converted into another unit impulse by the *impulse modulator* and applied to the second lag, resulting in a first sample of unity at the output.

Contrast this with the case when there is no impulse modulator. The input impulse will again produce an initial step at the output of the first lag, but this unity value will now only cause the second lag to ramp up with an initial slope of one unit per second. It is not until the following sample that any change in the output will be detected. All this is shown in Fig. P11-2.

The first important lesson is that when systems are connected in cascade, their z transform cannot in general be found by multiplying the separate z transforms—unless the subsystems are linked by an impulse modulator. Instead the time solution of the combined impulse response must be found, and the z transform calculated by summing the sampled series.

There is an even more important conclusion to be drawn from expression (P11-2). Suppose we change the continuous transfer function slightly, multiplying it by gain b to become

$$\frac{b}{(s + a)(s + b)} = \frac{1}{(1 + a)(1 + s/b)}$$

The second lag has a time constant $1/b$ and unity long-term gain. By letting b tend to infinity, this lag can be made negligibly short—at least in the continuous case. Now look at its effect on the z transform.

Multiplied by b, expression (P11-1) becomes

$$b \frac{e^{-a\tau} - e^{-b\tau}}{b - a} \frac{z^{-1}}{(1 - z^{-1} e^{-a\tau})(1 - z^{-1} e^{-b\tau})}$$

$$= \frac{e^{-a\tau} - e^{-b\tau}}{1 - a/b} \frac{z^{-1}}{(1 - z^{-1} e^{-a\tau})(1 - z^{-1} e^{-b\tau})}$$

If we let b tend to infinity, we might expect to arrive at expression (P11-1). We do not. Even when we have disposed of the second factor of the denominator and the exponential in $-b\tau$ in the numerator, we are left with

$$\frac{e^{-a\tau} z^{-1}}{1 - z^{-1} e^{-a\tau}} \qquad (\text{P11-3})$$

In fact, the output sample at $n = 0$ has 'gone missing', and the transform above corresponds to the sum of the rest of the series.

This is a much more realistic expression for the transform when used in analysing a computer control system. The computer will sample the relevant variables and then output a drive value. There is no way in which the output

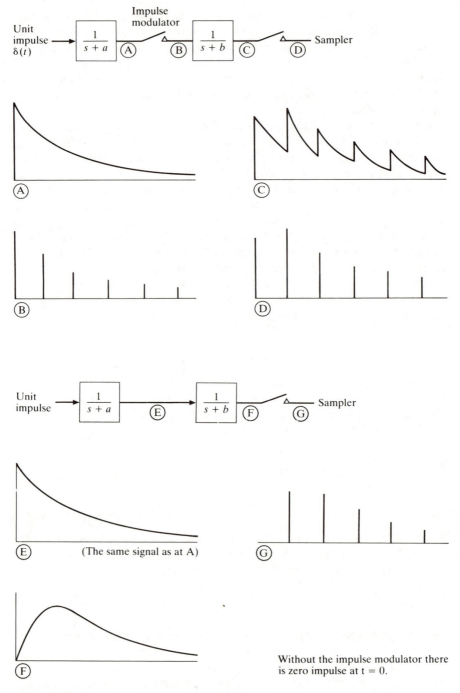

Figure P11-2 Effect of an impulse modulator in a cascaded system.

at $n = 0$ can influence the sample at $n = 0$, because it is applied 'just after' the sample is taken, $t = 0+$.

Do not be too alarmed by these philosophical problems. They arise from the notion of applying an impulse at the input to the system. A computer controller instead sets the finite value of an analogue output which remains constant until the next cycle. This *zero order hold* has quite different problems of its own, which are dealt with in the next chapter.

When examining continuous systems, we chose to consider outputs $\mathbf{y} = C\mathbf{x}$ which were a mixture of the state but not directly of the input. This neatly avoided the problem of 'algebraic' feedback, where there is an instantaneous loop from the input, through the system and back around the feedback path to the input. The actual value taken up by the input signal is then determined by simultaneous equations. In well-ordered computer control, this sort of feedback just cannot happen. A signal is sampled, another is output, and the order of the sampling and computation operations dictates whether one value can influence the other. In either event the output will steadily hold its value until the next time frame.

Exercise P11-3-1 *What are the z transforms corresponding to the following transfer functions?*
(a) *An integrator, $1/s$. (Argue the merits of two alternative answers.)*
(b) *An integrator combined with a lag, $1/s(s + a)$.*
(c) *An integrator followed by a time delay of 2.5 τ.*
(d) *Two integrators followed by a time delay of 2.5 τ.*
(e) *A time delay of 2.5 τ on its own.*

P11-4 THE BETA AND w TRANSFORMS

We met the beta operator in the prelude to the last chapter. As an approximation to differentiation it involved taking the 'next' value of the variable concerned minus its present value and dividing the result by the time between samples. It can be immediately expressed in terms of the z transform as

$$\beta = \frac{z - 1}{\tau}$$

If we have the equation $dy/dt = u$, we might represent it as a transform approximation

$$\beta Y = U$$

that is

$$\frac{z - 1}{\tau} Y = U$$

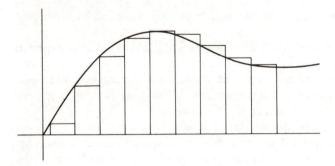

Figure P11-3 Euler rectangular integration.

This can be rearranged to represent the crude Euler integration performed by

$$y(n + 1) = y(n) + \tau u(n)$$

equivalent to integrating the area under a curve of u against time by taking a sum of rectangles, as shown in Fig. P11-3. In flowgraph form, the process is shown in Fig. P11-4. It can be represented by the software system:

```
Y = X
REM : The output is the total up to the last input
X = X + DT*U
REM : Now we update the state for next time
```

Instead of rectangular integration, we would much prefer to use trapezoidal integration, where the average input over the period is used (see Fig. P11-5). In other words, the amount that we must add to the output is half the sum of the present input and the next input

$$y(n + 1) = y(n) + \tau \frac{u(n) + u(n + 1)}{2} \tag{P11-4}$$

In transform terms, this becomes

$$\frac{z - 1}{\tau} Y = \frac{z + 1}{2} U$$

or

$$\frac{2}{\tau} \frac{z - 1}{z + 1} Y = U \tag{P11-5}$$

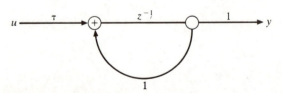

Figure P11-4 Flowgraph of Euler integration.

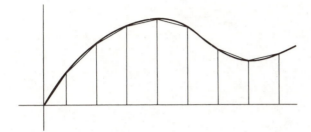

Figure P11-5 Trapezoidal integration.

We can introduce a new transform, the w transform, and write

$$wY = U$$

From the relationship

$$w = \frac{2}{\tau}\frac{z-1}{z+1}$$

we can solve for z, to find

$$z = \frac{2 + w\tau}{2 - w\tau}$$

This is a bilinear relationship; that is to say for each z there corresponds just one value of w and for each w there is just one z.

Now if you are alert, you will have seen that y is unsuitable as a state variable, as expressed in Eq. (P11-4). A variation in u produces an instantaneous change in y, whereas a state variable should carry forward information only of previous inputs. The output, y, must here be constructed from a mixture of the state and the input. Now

$$y(n + 1) = \tau\left[\frac{u(n + 1)}{2} + \frac{u(n)}{2}\right] + y(n)$$

$$= \tau\left[\frac{u(n + 1)}{2} + \frac{u(n)}{2} + \frac{u(n)}{2} + \frac{u(n - 1)}{2}\right] + y(n - 1)$$

$$= \tau\left[\frac{u(n + 1)}{2} + u(n) + u(n - 1) + \frac{u(n - 2)}{2}\right] + y(n - 2)$$

$$= \tau\left[u(n + 1)/2 + u(n) + u(n - 1) + \cdots + u(1) + \frac{u(0)}{2}\right] + y(0)$$

So the only real difference in the output compared with that of the Euler integrator is the addition of an instantaneous contribution $\tau u(n + 1)/2$. We

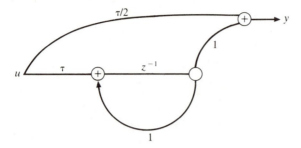

Figure P11-6 Trapezoidal integration flowgraph.

can realize the *trapezoidal integrator* by the flowgraph shown in Fig. P11-6, or by the software code:

```
Y = X + DT*U/2
REM : Y combines the old state with the new input
X = X + DT*U
REM : The state is updated for next time
```

There should be no problem in substituting w for s in the required transfer function and then implementing the corresponding discrete system by means of delay elements. There is some need for caution where feedback is concerned, since there will almost certainly be an 'algebraic' component. If the appropriate function of z is substituted for w, however, and the expression is tidied into the ratio of two polynomials in $1/z$, then any feedback algebra will almost certainly be straightened out in the process.

The question remains: how good an approximation will the new transform give to the representation of $sY = U$? The beta transform showed its shortcomings in Sec. P10-2, when the imaginary axis of the complex frequency plane was seen to map into a circle in the β plane. What does the w transform do to a complex sinusoid? We will see late in the next section that the imaginary frequency axis maps onto the imaginary w axis, while the real axis maps into the real w axis.

P11-5 MAPPINGS AND ALIASING

When we subject a complex sinusoid, $\exp(st)$, to the sampling process, we arrive at a sequence of values 1, $\exp(s\tau)$, $\exp(2s\tau)$ and so on. Each application of the operator z has the effect of multiplying the function by $\exp(s\tau)$. We can regard this relationship as defining a mapping from the complex frequency plane, in which s is defined, to the z plane (Fig. P11-8).

In Sec. P11-3 we saw that the lag $1/(s + a)$ corresponded to the z plane function

$$\frac{1}{1 - z^{-1} e^{-a\tau}} = \frac{z}{z - e^{-a\tau}}$$

The s plane pole at $-a$ has mapped into a z plane pole at $\exp(-a\tau)$. (The z plane zero at the origin can remain a matter for debate.)

Now when we take a point on the imaginary s axis, jω, it is seen to map into the z plane point

$$\cos(\omega\tau) + j \sin(\omega\tau)$$

As ω increases, this point sets out from $1 + j \cdot 0$ and moves anticlockwise around the unit circle. When $\omega = \pi/\tau$ the z locus crosses the real axis again at $-1 + j \cdot 0$, but at $\omega = 2\pi/\tau$ it is back at a value of unity. How can this be? If we sample a sine wave at intervals of exactly one cycle, all our samples will be equal; we cannot tell whether we have a sine wave or a constant value. It gets worse. If the sine wave is of a slightly higher frequency, the samples show a sinusoidal behaviour, but appear to arise from a low frequency. This phenomenon, *aliasing*, is illustrated in Fig. P11-7.

In fact the imaginary axis of the frequency plane is mapped over and over again onto the unit circle, and the digital system has no means of resolving how many multiples of $2\pi/\tau$ should be added or subtracted from the apparent frequency of any given signal. It is usual to add analogue low-pass

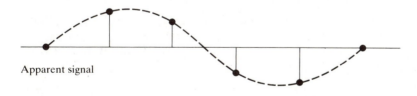

Apparent signal

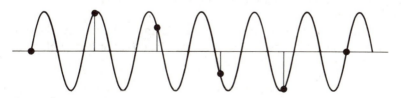

True signal

Figure P11-7 An illustration of aliasing. The sampled signal may falsely appear to be of a lower frequency.

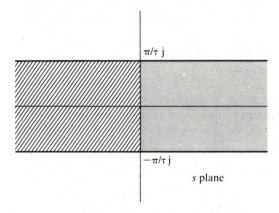

$\pi/\tau\, j$

$-\pi/\tau\, j$

s plane

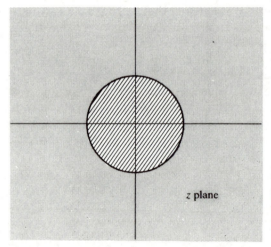

z plane

Figure P11-8 Mapping of the *s* plane to the *z* plane. Each horizontal 'stripe' of the *s* plane maps on to the *z* plane, as shown.

antialiasing filters before an analogue-to-digital converter for the express purpose of preventing this effect, by severely attenuating any frequency components above π/τ.

Now we can look at the mapping from the complex frequency plane to the *w* plane to see how well poles and zeros can be placed in the *w* plane to represent poles and zeros in the *s* plane:

$$\mathcal{W}(e^{st}) = \frac{2}{\tau}\frac{e^{s\tau} - 1}{e^{s\tau} + 1}\,e^{st}$$

$$= \frac{2}{\tau}\tanh\left(\frac{s\tau}{2}\right)e^{st}$$

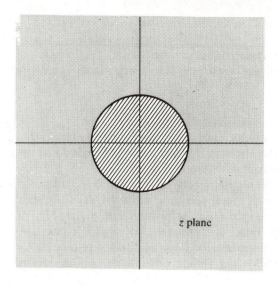

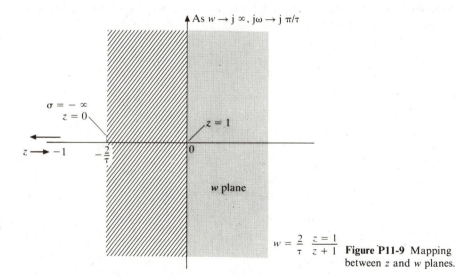

$$w = \frac{2}{\tau} \frac{z = 1}{z + 1}$$

Figure P11-9 Mapping between z and w planes.

If s is imaginary, with value $j\omega$, then the corresponding w plane point is

$$\frac{2}{\tau} j \tan\left(\frac{\omega\tau}{2}\right)$$

For frequencies much less than $2/\tau$ the w plane point will correspond closely with s. As the frequency increases, however, w will rush off along the imaginary axis approaching infinity as ω tends to π/τ. Poles on the imaginary

s axis correspond to poles on the imaginary w axis, even if they become 'warped' at higher frequencies. To place them accurately at a specified frequency, they can be 'prewarped' in compensation.

Real s poles will also suffer warping if too large, from value a to value

$$\frac{2}{\tau} \tanh\left(\frac{a\tau}{2}\right)$$

The positive real s axis is mapped just once into the positive real w axis, from $w = 0$ to $w = 2/\tau$; similarly the negative real s axis is mapped just once onto the negative real w axis from $w = 0$ to $w = -2/\tau$.

When we consider the mapping between the z and w planes, we see the interior of the unit circle of z mapped once onto the negative real w half-plane, and the exterior mapped onto the positive half-plane (Fig. P11-9).

We now have two lines of approach for digital control. We can compensate a system by direct consideration of its discrete-time behaviour or we can translate controllers which should work in continuous time into their discrete-time equivalents.

Exercise P11-5-1 *In a system sampled ten times per second, what discrete-time controller is equivalent to the phase advance* $(1 + 2s)/(2 + s)$?

ELEVEN

DESIGN METHODS FOR COMPUTER CONTROL

11-1 INTRODUCTION

In the last chapter, we saw that the state-space methods of continuous time could be put to good use in a discrete-time control system. Now we look at techniques involving transforms and the frequency domain, and see that these too have their counterparts. Some design methods seem straightforward, such as the root locus, while others can conceal pitfalls. When continuous and discrete feedback are mixed, analysis can be particularly difficult.

11-2 THE DIGITAL-TO-ANALOGUE CONVERTER AS 'ZERO-ORDER HOLD'

In any practical digital control system, the result of the computations must at some point be output to the 'real world', to the system being controlled. The output number is not converted into an infinite impulse, as the Laplace transform would have it, but its value is lodged in an electronic circuit from one control cycle to the next. The number is translated into a signal which is applied to the process and hopefully controls it. The translation circuit is a digital-to-analogue converter.

The nature of the converter depends very much on the application. At the simplest level it may consist of no more than a relay or valve which delivers power to the heater or actuator. Another application may require the output

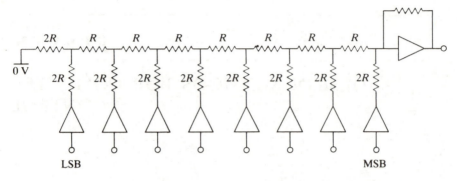

Figure 11-1 Workings of a ladder type eight-bit DAC. Low-cost devices contain data latches to interface directly to the computer bus.

to be a current accurate to one part in many thousands, perhaps to deflect an electron beam in a silicon fabrication process. For now we will look at the substantial middle ground, where a conversion accuracy of one part in 256 is more than sufficient.

A system such as a position controller may require the motor drive to have several possible levels, since switching the power from one extreme to the other could give jerky and unacceptable performance. On the other hand, there is little point in seeking great accuracy and resolution in the drive value, since it is surrounded by a tight position feedback loop. Eight-bit digital-to-analogue converters (DACs) are the most common. They are of low cost, since the manufacturing tolerance is easy to meet. Most control computers are equipped to output eight-bit bytes of data, so there is little incentive to try to use a simpler DAC. For interest, an eight-bit DAC is illustrated in Fig. 11-1.

The *quantization* of a DAC means that a compromise has been made to round the value which should have been output to the nearest DAC value available. When 256 levels define all the output values from full drive forwards to full drive reverse, each step represents nearly one per cent of the drive amplitude. For now let us ignore this source of error and concentrate on the time behaviour.

Suppose that we have a motor described in Laplace terms as $1/s(s + a)$. How do we get from the notional impulses within the discrete-time part of the system to the continuous-time analogue section? The DAC takes the mathematical form of a digital filter driving an integrator.

At each moment of output, the DAC output may make a step change. This step is the integral of an impulse. If we have a sequence of outputs

$$1 \quad 4 \quad 8 \quad 5 \quad 3 \quad 2 \quad 6 \quad \cdots$$

then these can be regarded as the output of an integrator driven by a

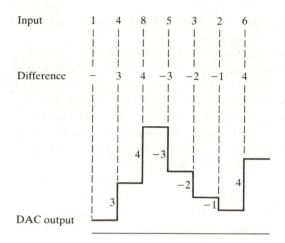

Input   1   4   8   5   3   2   6

Difference   —   3   4   −3   −2   −1   4

DAC output

Input ⟶ $1 - z^{-1}$ ⟶ Difference ⟶ $1/s$ ⟶ DAC output

Figure 11-2 Hybrid z and s representation of a DAC.

sequence of impulses with values

$$? \quad 3 \quad 4 \quad -3 \quad -2 \quad -1 \quad 4 \quad \cdots$$

The impulses are the differences between the latest value of output and the previous one. You will note that we cannot specify the value of the first such impulse without knowing the DAC level before it.

The business of taking the difference between output values is very simply performed by the digital filter

$$1 - z^{-1}$$

while the integrator is the Laplace function $1/s$. The result is a hybrid (Fig. 11-2):

$$(1 - z^{-1})\frac{1}{s}$$

The implication is that when making a z transform analysis of the system, the $1/s$ term must be incorporated in the analogue transfer function before the z transform equivalent is found. (Remember that the z transform equivalent of the product of Laplace transfer functions is very rarely equal to the product of the individual z transforms.)

In the case of the position controller, we must find the z transform equivalent not of $1/s(s + a)$ but of $1/s^2(s + a)$—although if we also have a tacho signal then this will have another transfer function of its own. This representation of a DAC is termed a *zero-order hold*.

There is need for a word or two of caution. Buried within the integrator is an extra state. The integrator output is the cumulative sum of differences between successive inputs to the DAC. If this sum should become misaligned with the input sequence, there is no mechanism to pull it back into step. On the other hand, since the integrator is a conceptual one there is also no reason for it to get out of step. The extra state produces an extra denominator term $(1 - 1/z)$ in the z transform equivalent of the system, which will usually cancel with the zero at $z = 1$ from the differencing function.

Exercise 11-2-1 *What is the z transform equivalent of a motor described by $s^2 Y = U$, when driven by a DAC updated ten times per second? Can the motor be controlled stably by position feedback alone?*

Exercise 11-2-2 *Can the motor position be stabilized using the digital phase advance of Exercise P11-5-1?*

11-3 A POSITION CONTROL EXAMPLE—DISCRETE-TIME ROOT LOCUS

With the added integrator, the analogue part of the system of Exercise 11-2-1 is described by

$$Y(s) = \frac{1}{s^2} U(s)$$

The equivalent z transform can be looked up in a table of transform equivalents, but it is more instructive to see ways of working it out. The impulse response is the time function

$$y(t) = \frac{t^2}{2}$$

which when sampled at times $n\tau$ becomes

$$y(n) = n^2 \frac{\tau^2}{2}$$

Let us develop a general rule for extending a table of transforms to include powers of n. If

$$X(z) = \sum_{n=0}^{\infty} x(n)z^{-n}$$

then

$$\frac{dX}{dz} = \sum_{n=0}^{\infty} x(n)(-n)z^{-(n+1)}$$

Turning this around and multiplying by $-z$ gives us

$$Z\{nx(n)\} = -z \frac{dX}{dz}(z) \tag{11-1}$$

We know that the transform of the sequence $1, 1, 1, \ldots$, where $x(n) = 1$, is $1/(1 - 1/z)$, so to find the transform of $x(n) = n^2\tau^2/2$ we apply rule (11-1) twice and multiply by $\tau^2/2$ to arrive at

$$Z\left\{n^2 \frac{\tau^2}{2}\right\} = \frac{\tau^2}{2} z \frac{d}{dz}\left[z \frac{d}{dz}\left(\frac{z}{z-1}\right)\right]$$

$$= \frac{\tau^2}{2} \frac{z(z+1)}{(z-1)^3}$$

We must now include the differencing function to obtain the transfer function from the DAC to the position sampler:

$$G(z) = \frac{\tau^2}{2} \frac{1-z}{z} \frac{z(z+1)}{(z-1)^3}$$

$$= \frac{\tau^2}{2} \frac{z+1}{(z-1)^2} \tag{11-2}$$

We should have been able to obtain the same result by calculating the transfer function matrix from the matrix state equations. This same system was investigated in Sec. 10-6, and its discrete state equation was expressed in (10-19). If we take the z transform of the general discrete state equations they change slightly to become

$$z\mathbf{X} = A\mathbf{X} + B\mathbf{U}$$

$$\mathbf{Y} = C\mathbf{X}$$

The first of these can be rearranged as

$$(zI - A)\mathbf{X} = B\mathbf{U}$$

to give

$$\mathbf{X} = (zI - A)^{-1}B\mathbf{U}$$

and

$$\mathbf{Y} = C(zI - A)^{-1}B\mathbf{U} \tag{11-3}$$

Exercise 11-3-1 Show that when C is equated to $(1 \quad 0)$ to select the position as output, substituting the values of the matrices of (10-9) into expression (11-3) gives the same transfer function as (11-2).

Now if we close the loop, using a gain k at the input as shown in Fig. 11-3, we have a closed loop gain

$$\frac{kG(z)}{1 + kG(z)}$$

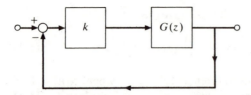

Figure 11-3 Unity feedback around a system with variable gain.

We can look at the poles algebraically by multiplying out the denominator, obtaining

$$(z - 1)^2 + k\frac{\tau^2}{2}(z + 1) = z^2 + z\left(-2 + k\frac{\tau^2}{2}\right) + \left(1 + k\frac{\tau^2}{2}\right)$$

This is virtually the same quadratic as the one found in Sec. 10-6, leading to the conclusion that proportional control alone was always unstable. Since we have an adjustable parameter, k, could we learn more from the locus of the poles as k is varied? Could we investigate the root locus?

We have two poles at $z = 1$ and a zero at $z = -1$. The rules at the end of Section 7-5 give us the following clues for sketching the root locus:

1. For large k, a single 'excess pole' makes off to infinity along the negative real axis.

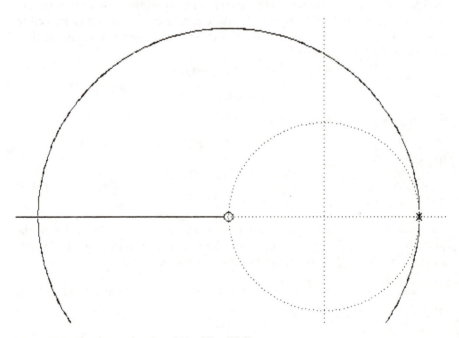

Figure 11-4 Root locus of system $G(z)$ of Eq. (11-2).

2. Only that part of the real axis to the left of the zero can form part of the plot for positive k (i.e. for negative feedback).
3. The derivative of $G(z)$ is zero at $z = -3$, a point that lies on the locus (an exercise for the reader). This point is therefore a junction of branches of the locus.

We deduce that the locus splits and leaves the pole pair 'north and south', following a pair of curved paths that come together again at $z = -3$. One branch now follows the real axis to the zero at $z = -1$, while the other travels along the axis in the opposite direction to minus infinity. The curved part of the locus is in fact a circle, and the locus is illustrated in Fig. 11-4. It is plain that the locus lies entirely in the forbidden region outside the unit circle, so that stability is impossible.

By adding a dynamic controller with further poles and zeros to the loop, can we 'bend' the locus to achieve stability?

11-4 DISCRETE-TIME DYNAMIC CONTROL—ASSESSING PERFORMANCE

Can we stabilize the position system with the digital equivalent of a phase advance? Suppose we take the transfer function $(1 + 2s)/(2 + s)$ suggested in Exercise P11-5-1 and use the w transform to find an approximate equivalent. Then we substitute w for s, where

$$w = \frac{2\,z - 1}{\tau\,z + 1}$$

to get a transfer function

$$\frac{1 + \dfrac{4\,z - 1}{\tau\,z + 1}}{2 + \dfrac{2\,z - 1}{\tau\,z + 1}} = \frac{\tau\,(z + 1) + 4(z + 1)}{2\tau\,(z + 1) + 2(z - 1)}$$

(substitute $\tau = 0.1$ seconds)

$$= \frac{4.1z - 3.9}{2.2z - 1.8}$$

$$= 1.86\,\frac{z - 0.951}{z - 0.818}$$

A zero at $z = 0.951$ and a pole at $z = 0.818$ are added to the root locus diagram, with a result illustrated in the plots shown in Fig. 11-5. Stable control is now possible, but the settling time is not short. How can we tell?

In the negative real half of the s plane, the real part of s defines an

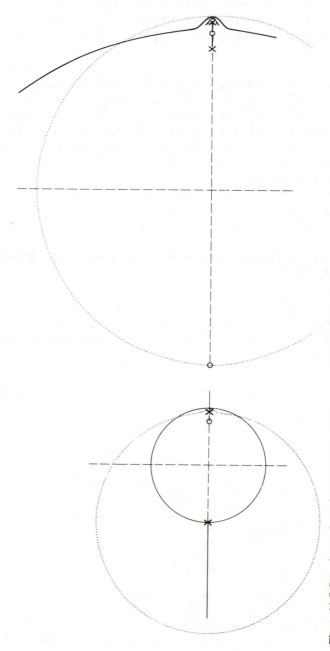

Figure 11-5 Root locus plots of compensated system—general and fine detail.

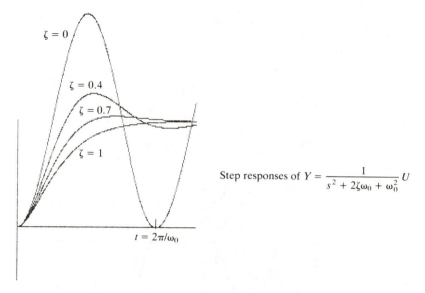

$$\text{Step responses of } Y = \frac{1}{s^2 + 2\zeta\omega_0 + \omega_0^2} U$$

Figure 11-6 Second-order responses for varying damping factor.

exponential decay rate while the imaginary part defines a frequency. A complex pair, $-\sigma \pm j\omega$, can be viewed as the roots of a quadratic

$$s^2 + 2\sigma s + \sigma^2 + \omega^2 = 0$$

Such a quadratic is alternatively expressed in terms of the *undamped natural frequency* ω_0 and damping factor ζ as

$$s^2 + 2\zeta\omega_0 + \omega_0^2 = 0$$

The damping factor dictates the shape of the step response while the natural frequency determines its speed. A collection of responses for varying values of ζ are shown in Fig. 11-6.

Now ζ is tied to σ and ω by the relationship

$$\zeta = \frac{\sigma}{\sqrt{\sigma^2 + \omega^2}}$$

In particular, if $\sigma = \omega$ then $\zeta = 0.7$, which gives a response of acceptable appearance. This is a very popular choice of damping factor, for no more complicated reason than that the poles lie on lines drawn at 45° through the s plane origin! The natural frequency, the frequency at which the system would oscillate if the damping were removed, is now represented by the distance from the origin. Figure 11-7 illustrates the situation.

To translate these yardsticks into the discrete-time world, we must use the relationship $z = \exp(s\tau)$. Damping factors and natural frequencies are mapped into the interior of the unit circle as shown in Fig. 11-8.

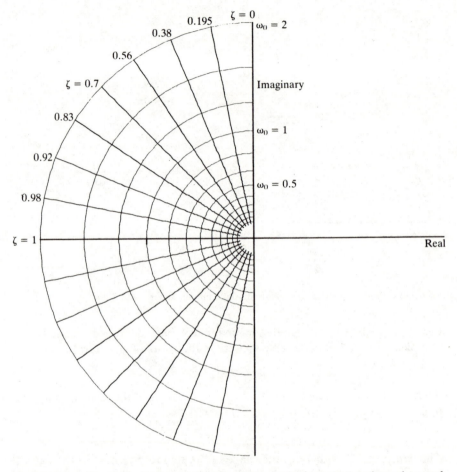

Figure 11-7 s plane loci of constant damping factor and of constant undamped natural frequency.

We are now in a position to judge how effective a compensator can be by examining its root locus. We can clearly do better than the gentle phase advance we started with. We could perhaps mix the output with a *derivative* formed by subtracting its previous value

$$u = ay(n) + [y(n) - y(n - 1)]$$

that is

$$U = \left[(a + b) - \frac{b}{z} \right] Y$$

$$= \frac{(a + b)z - b}{z} Y$$

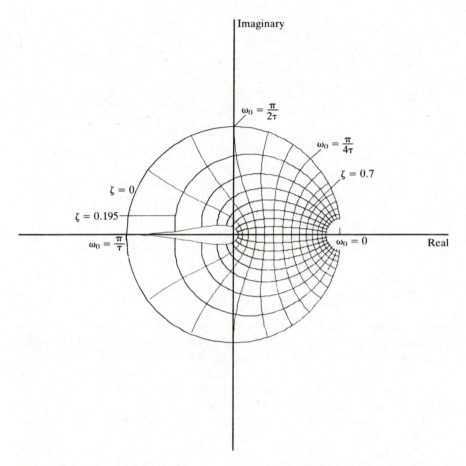

Figure 11-8 Corresponding z plane loci of constant damping factor and of constant undamped natural frequency.

This gives a pole at the origin and a zero which we can place anywhere on the segment of the real axis joining the origin to $z = 1$. A root locus for the case where the zero is at $z = 0.5$ is shown in Fig. 11-9. In software it would be implemented for some chosen value of gain by

```
U     = K*(2*Y − YOLD)
YOLD = Y
```

In Sec. 10-6 a controller was devised to give a dead-beat response. Expression (10-14) shows that it has a pole at a negative real value of z, at $z = -0.75$, giving an even more dramatic modification of the root locus. The zero is now at $z = 0.6$ and the result is illustrated in Fig. 11-10.

Could the system have been more easily deduced from its s plane transfer

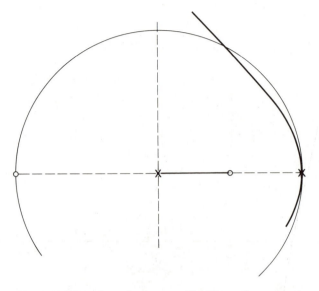

Figure 11-9 Root locus for system with 'difference' compensator.

function, using the approximate equivalence of s to w? If we substitute w for s and then replace w by $2(z - 1)/\tau(z + 1)$ we will have three zeros at $z = 1$ and three poles at $z = -1$. Adding in the 'differencer' removes one of the poles, but adds another pole at $z = 0$. The result bears little resemblance to the system we must actually control. The w transform may enable us to devise filters, but could be misleading for modelling a system.

On the other hand, the z transform transfer function can be translated

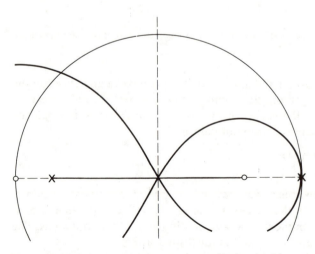

Figure 11-10 Root lcous for compensator achieving dead-beat performance.

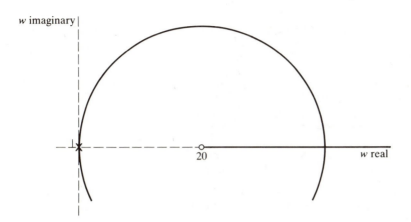

w imaginary

20   *w* real

Figure 11-11 *w* plane root locus of uncompensated system. There are two poles at the origin and a zero at $+20$.

exactly into the *w* plane by substituting $(2 + w\tau)/(2 - w\tau)$ for *z*. Now a *w* plane compensator can be tried out, plotting the resulting root locus in the *w* plane. The stability criterion is the familiar one of requiring all the poles to have negative real parts.

Exercise 11-4-1 *Represent the motor position problem by a transfer function in w; sketch its root locus. What is the effect of a compensator with transfer function (1 + w)? Can such a compensator be realized?*

Making the substitution gives us a transfer function $(1 - w\tau/2)w^2$ for the DAC and motor. With two poles at the origin and a zero at $w = 20$, the root locus illustrated in Fig. 11-11 shows that feedback alone will not suffice. The compensator adds a second zero at $w = -1$. Do the twin poles now split, one moving to the left and the other to the right along the real axis to each zero? Not a bit.

The transfer function has an extra negative sign, seen when it is rewritten as $-(w\tau/2 - 1)w^2$, so the valid real-axis part of the locus lies outside the zeros, not between them. The poles split north and south, following a circular path to recombine on the negative real axis. One pole now approaches the zero at $w = -1$, while the other makes off to minus infinity, reappearing at plus infinity to return along the real axis to the second zero. Figure 11-12 shows the result.

The compensator is not hard to realize. Another substitution gives us the transfer function $k(21z - 19)/(z - 1)$, which can be rewritten as

$$U = k \frac{21 - 19z^{-1}}{1 - z^{-1}} Y$$

or

$$U = -z^{-1}U + k(21 - 19z^{-1})Y$$

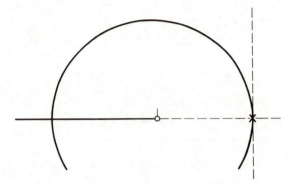

Figure 11-12 *w* plane root locus with (1 + *w*) compensator.

This will be realized by the lines of software:

```
U     = −UOLD + K*(21*Y − 19*YOLD)
UOLD = U
YOLD = Y
```

When preceded by an input statement to read *Y* from the sampler and followed by a command to output *U* to the DAC, this will construct the compensator in question. The compensator looks unstable in isolation, but when incorporated in the loop should be able to perform as predicted.

It appears deceptively simple to place a few compensator poles and zeros to achieve a desired response. There must be further objectives and criteria involved in designing a control system. Indeed there are. Our problems are just beginning.

11-5 ERRORS AND DISTURBANCES

Until now, our systems have included inputs, states and outputs. We have considered initial disturbances and have assumed that the goal of the controller is to bring the error to zero in the absence of any other input. To see that this is an oversimplification, consider a simple water-heater experiment (Fig. 11-13).

Water flows over a heating element and via a pipe to the output, where its temperature is measured. An input knob sets the target temperature and power is supplied to the heater in proportion to the output error. If the control is stable, then in the steady state

$$T_{\text{out}} = Au$$

and

$$u = k(T_{\text{demanded}} - T_{\text{out}})$$

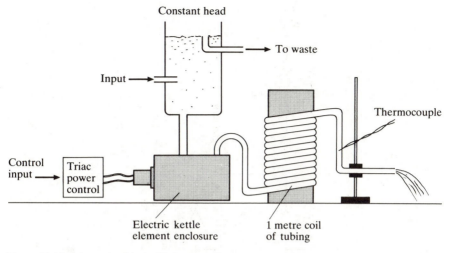

Figure 11-13 Schematic of water-heater experiment.

where the temperatures are measured as the difference from ambient. Algebra shows that

$$T_{\text{out}} = T_{\text{demanded}} \frac{kA}{1 + kA}$$

For any setting above ambient, there must always be a non-zero error to ensure that the heater supplies power. Unless kA is large, the temperature will not accurately follow the demand.

In this experiment, devised by the author many years ago for the Cambridge University Engineering Department, the pipe is long enough to introduce a delay of several seconds. For any but a modest value of kA, the system becomes unstable. Its Whiteley plot is a 'Swiss-roll' spiral which imposes a serious limit on the maximum allowable loop gain. Some additional method must be used to reduce the error, if accurate control is needed. Since the flow may vary which in turn will affect the value of A, recalibrating the demand knob will not suffice!

The accepted solution is to use integral control (see Fig. 11-14). In addition to the proportional feedback term, an integrator is driven by the error signal and its output is added to the drive, changing slowly enough that stability is not lost. The eventual error will now be reduced to zero, since the integrator can 'take the strain' of maintaining a drive signal. The method has its drawbacks. The integrator has increased the order of the system, which does not help the problem of ensuring stability. There is also the problem of *integral wind-up*.

Suppose that after the system has settled, the demand is suddenly

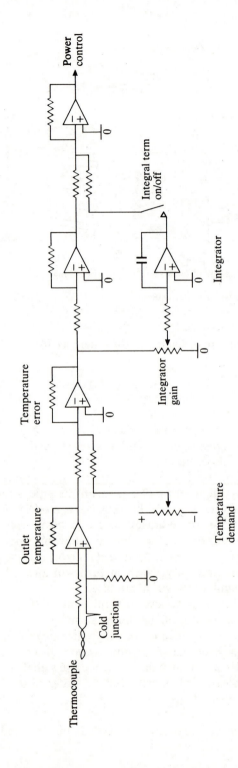

Figure 11-14 Controller with integrator.

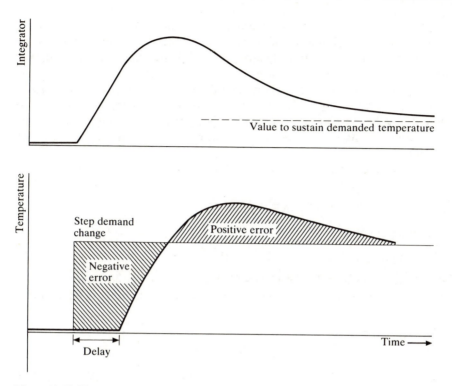

Figure 11-15 Time response showing 'integral wind-up'.

increased as shown in Fig. 11-15. The integrator value will be driven upwards by the error all the time the system warms to catch up, and will reach a value well in excess of that needed to sustain the new target temperature. The temperature will then overshoot, so that the error–time curve can have an area of excess temperature nearly equal to the area of the error time when warming.

Integral wind-up can be held in check by ensuring that the output of the integrator limits at a level only slightly in excess of the greatest value needed. An analogue limiter can also be applied at the input of the integrator to reduce the wind-up effect of a large error. In a digital controller, more sophisticated algorithms can ensure that integration does not start until the rate of change of output has fallen below a certain level. The algorithm can be tailored to the physical needs of the system.

Many principles of controller design have arisen from the needs of gunnery. A controller containing one integration can reduce to zero the aiming error with a stationary target, but will have a tracking error proportional to the target's velocity. With a second integrator we can track a target without error at constant velocity, but aim will be thrown off by acceleration. Each additional integrator raises the order of the polynomial that can be followed, but introduces more prolonged transient behaviour in acquiring the target. Fortunately most industrial problems can be solved with no more than a single integration.

A controller most commonly found in process control is the 'PID' controller—proportional plus integral plus derivative. Proportional control should now be familiar, integral has just been discussed and derivative is no more than the phase advance contribution which 'observes' the derivative of the output.

The simplest implementation of integral control is to compute an error, such as

TEMPERR = TEMPDEMAND − TEMPOUT

and 'integrate' it with

INTEGRAL = INTEGRAL + TEMPERR

after which a term **INTEGRAL∗INTGAIN** is added to the computation of the drive, U. To improve the practical performance, a number of 'IF's can be added, for example

IF ABS(TEMPERR) > LIMIT THEN
TEMPERR = LIMIT∗SGN(TEMPERR)

will put a limit on the wind-up effects described above. Intuitively this appears sensible, but it has the drawback that a non-linearity has now been introduced into the system. Performance may be improved, but the task of analysing that performance becomes more difficult. Simulation will still be effective, but algebraic methods may only give an approximation.

Integral control is a method open to a simple analogue controller. When the controller is digital, many more sophisticated techniques are available. The controller is 'aware' of a change in demand, and integral correction can pause until the response has settled. The error can then be integrated over a finite, predefined interval of time to find the average correction needed, and this can be applied as a step change of input. After a delay to allow the transient to pass, the integration can be repeated. In this way, there is a hope of achieving dead-beat correction of the offset, instead of the slow exponential decay resulting from continuous integral feedback.

Now we must investigate the various forms of disturbance in a more general way with the aid of some block diagrams (Fig. 11-16):

N1. The command input is subject to noise of various forms, from quantization of a numeric command to tolerances in encoding a control lever. This error is outside the loop and therefore nothing can be done about it. The target will be the signal as interpreted by the control electronics.

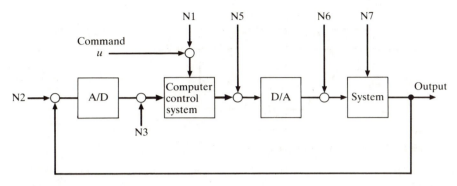

Figure 11-16 Block diagram of system, illustrating sources of noise.

N2. An error in measuring the value of the output for feedback purposes is equally immune from compensation. It is the sensor signal which the loop corrects, rather than the true output. If a position transducer has slipped 10°, then the controller will steadfastly control the output to a position with a 10° error.

N3. This represents the quantization error of the feedback digitizer. N2 arises by accident; N3 is deliberately permitted by the system designer when selecting the fineness of the digitization, perhaps as the bit-length of an analogue-to-digital converter (ADC).

N4. The computation of the algorithm will not be perfect, and extra disturbances can be introduced through rounding error, shortened multiplications and approximated trigonometric functions.

N5. Another portion of quantization error is served up when driving the DAC output. The magnitude of this error is determined by the system designer when selecting the DAC word length.

N6. The DAC is a fallible electronic device, and unless carefully set up can be subject to offset error, to gain error and to bit-sensitive irregularities.

N7. This is the noise disturbing the system that the controller has really been built to combat. It can range from turbulent buffeting in an aircraft to the passenger load in an elevator, from the switching of an additional load onto a stabilized power supply to the opening of a refrigerator door.

In a linear system, the noise source N7 can be replaced by a transformed signal $N(s)$ representing its effect at the output of the system. The equations of the feedback system shown in Fig. 11-17 become

$$Y(s) = G(s)U(s) + N(s)$$

$$U(s) = F(s)V(s) - H(s)Y(s)$$

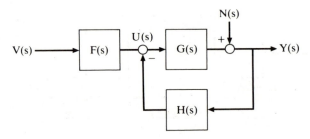

Figure 11-17 Control loop with disturbance noise.

Now

$$Y(s) = G(s)[F(s)V(s) - H(s)Y(s)] + N(s)$$

$$[1 + G(s)H(s)]Y(s) = G(s)F(s)V(s) + N(s)$$

or

$$Y(s) = \frac{G(s)F(s)}{1 + G(s)H(s)} V(s) + \frac{1}{1 + G(s)H(s)} N(s)$$

Our aim must be to minimize the effect of the disturbance, $N(s)$. If $V(s)$ is a command representing the desired value of the output, then we also want the first transfer function to be as near to unity as possible. In the tracking situation, the system may be simplified and redrawn as shown in Fig. 11-18. We see that

$$Y(s) = \frac{G(s)H(s)}{1 + G(s)H(s)} V(s) + \frac{1}{1 + G(s)H(s)} N(s)$$

$$= V(s) + \frac{1}{1 + G(s)H(s)} [N(s) - V(s)]$$

To establish the ability of the system to track perturbations of various kinds, we can substitute functions for $N(s)$ representing a step, a ramp or a higher power of time. Then we can use the final value theorem to find the ultimate value of the error.

Exercise 11-5-1 Show that with an $H(s)$ of unity the system $1/s(s + 1)$ can track a step without final error, but that $1/(s + 1)$ cannot.

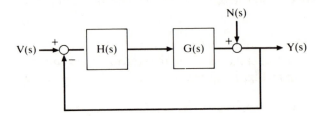

Figure 11-18 Tracking system with noise.

Exercise 11-5-2 *Show that by adding integral control, with $H(s) = 1 + a/s$, the system $1/(s + 1)$ can be made to track a step with zero final error.*

The use of large gains will impose heavy demands on the drive system; for any substantial disturbance the drive will limit. Large derivative or difference terms can also increase the drive requirements, made worse by persistent high frequency noise in the system. If the drive is made to rattle between alternative limiting extremes, the effect is to reduce the positional loop gain so that offset forces will result in unexpectedly large errors.

The pursuit of rapid settling times will have a penalty when noise is present or when demand changes are likely to be large. Linear design techniques can be used for stabilization of the final output, but many rapid-response positioning systems have a separate algorithm for handling a large change, bringing the system without overshoot to the region where proportional control will be effective. As we saw in Sec. 3-5, the design of a limiting controller requires the consideration of many other matters than eigenvalues and linear theory.

Exercise 11-5-3 *Although daisywheel printers (Fig. 11-19) are out of fashion, the servomotor systems for positioning the daisy present some interesting design*

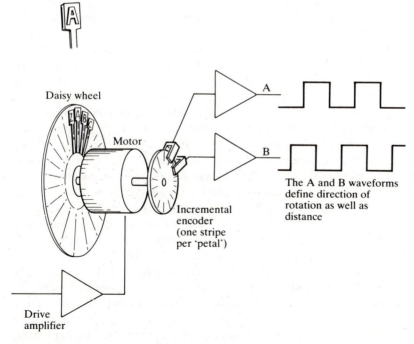

Figure 11-19 Schematic of daisywheel controller.

problems and possibilities. Suppose that photocells generate two digital waveforms in quadrature, as used in the position experiment of Chapter 4, where there is one cycle per 'petal'. A further signal identifies the 'home' petal. A d.c. motor completes the mechanical system. Can adequate performance be achieved without additional sensing?

11-6 PRACTICAL DESIGN CONSIDERATIONS

We often find that the input to an analogue-to-digital converter is contaminated by the addition of noise. If the noise is of high frequency it may be only necessary to use a low-pass filter to clean up the signal. We have already seen that the one-line computation

$$X = X + (SIGNAL - X)/K$$

will approximate to a filter with time constant K times the sampling interval, so will this suffice? Unfortunately we are caught out by aliasing. Whereas an analogue low-pass filter will attenuate the signal more and more as the frequency increases, the digital filter has a minimum gain of $1/(2K - 1)$. Indeed, as the input frequency increases towards the sampling frequency the gain climbs again to a value of unity (see Fig. 11-20).

We cannot use digital filtering, but must place an analogue low-pass filter before the input of the ADC. This will probably provide a second benefit. An ADC has a limited input range, corresponding to its, say, 256 possible answers. For efficient use of the conversion, the input signal range should nearly 'fill' that of the ADC. If allowance has to be made in the range for the presence of noise on top of the signal, a smaller signal amplitude must be used and the quantization noise of range/256 becomes more important.

There is, however, a situation where digital filtering can be more effective than analogue. Signals in industrial environments, such as those from strain gauges, are often contaminated with noise at the supply frequency, 50 or 60 Hz. To attenuate the supply frequency with a simple filter calls for a *break frequency* of the order of one second; this may also destroy the information we are looking for. Instead we can synchronize the converter with the supply frequency, taking two or four (or more) readings per cycle. Now taking the sum of the last two or four readings will cause the supply frequency contribution to cancel—although harmonics may be left. With twelve readings per cycle, all harmonics can be removed up to the fifth.

It may be that the input signal contains frequencies well above the supply frequency, so that even the filtering suggested above would obscure the required detail. It is possible to sample at a much higher multiple of the supply frequency and to update an array that builds an average from cycle to cycle of the measurement at each particular point in the cycle. In other words, the array reconstructs the cyclic noise waveform in which the non-cyclic

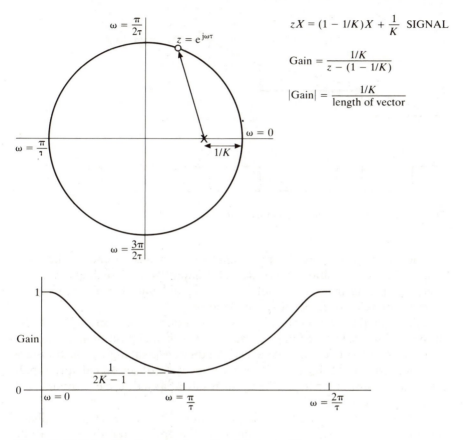

Figure 11-20 Gain versus frequency for a discrete low-pass filter.

signal is averaged out except for its steady d.c. term. The average d.c. of the array can be calculated and removed, so that the remaining element of the observed noise waveform can be used to compensate each sample as soon as it is taken.

Care is needed when selecting an ADC for the feedback signal. It is obvious that the quantization error must be smaller than the required accuracy, but there are other considerations. If the control algorithm depends on differential action of any form, the quantization effect can be aggravated. If sampling is rapid, so that there are many control cycles between changes in value of the digitized signal, then the difference between subsequent samples will appear as sporadic unit pulses.

Even when a higher resolution ADC is used, as hinted in Fig. 11-21, velocity control is likely to be coarse unless the resolution and cycle times are carefully balanced.

When the open loop system is not in itself well damped, it often makes

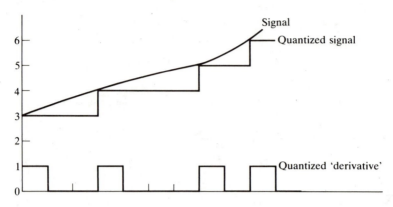

Figure 11-21 A signal is quantized and then differences are taken.

the control task much easier to apply analogue feedback around a tight 'inner loop'. Thus the position controller of Chapter 4 has analogue velocity feedback, permitting the output of the digital controller to represent a demand for velocity rather than acceleration.

The quantization problem shows up again in the example of the daisywheel servo, featured in Exercise 11-5-3. The final position must be held to a very small fraction of a 'petal', so the quarter-petal logic transitions are quite inadequate. Moreover, without a velocity feedback signal, any rapid-response control algorithm is likely to result in a limit cycle several signal transitions in amplitude.

The solution is to use a mixed-mode controller, with analogue feedback holding the final position. The signal from the optical transducer has to be 'squared up' to provide a logic signal. In its 'raw' state it varies continuously about a mid-level, and can readily be used with phase advance for analogue position control over a range from the trough to the peak of its waveform. The digital control mode now has the simpler task of bringing the position error and velocity sufficiently near to zero that the analogue control can lock in. This can be achieved in a highly non-linear manner to minimize the settling time (more about this in the next chapter).

The analysis of a system with an analogue inner loop is relatively straightforward, although some design decisions are now separated and more difficult to make. The continuous system can be expressed in terms of Laplace operators, and root locus, block diagram or algebraic methods can help decide on the feedback details. This entire block can then be transformed back into the time domain and sampled, so that a z transform representation can be made for the path from DAC at the system input to ADC sensing the output. The command input is also in sampled form, so that it can be represented by a z transform, and the closed loop transfer function can be calculated as the ratio of the z transform of the output divided by the z

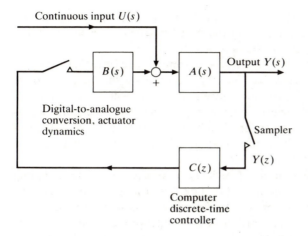

Continuous input $U(s)$

$B(s)$

$A(s)$

Output $Y(s)$

+

Digital-to-analogue
conversion, actuator
dynamics

Sampler

$C(z)$

$Y(z)$

Computer
discrete-time
controller

$$Y(z) = \frac{Z\{A(s)U(s)\}}{1 - C(z)Z\{A(s)B(s)\}}$$

It is not possible to express $Y(z)$ in the form $Y(z) = G(z)U(z)$

Figure 11-22 It can be impossible to derive a transfer function if a system has both discrete and continuous feedback.

transform of the input. Now the digital control and compensation can be designed to place the closed loop z poles to the designer's choice.

If the command input arrives in analogue form, and rather than being sampled by the controller is applied directly to the system, then it is another story. Just as the z transform of two cascaded analogue filters is unlikely to be the product of their individual transforms, so the z transform of the output will not be the z transform of the command input multiplied by the z transform of the system response. If a digital outer loop is applied, say to apply integral control, then the system cannot really be analysed in z transform terms. Instead, the digital part of the system will probably be approximated to its continuous equivalent, allowing design to be performed in the s plane.

The choice of sampling rate can be important, particularly when the system contains a pure time delay associated with an otherwise stable system. I am sure that you are familiar with the task of adjusting the temperature of a bathroom shower—the sort that has a length of pipe between the mixer taps and the shower head. The successful strategy is to try a setting, then have the patience to wait until the new mixture has passed through the system before making a compensating adjustment. If you are too hasty, you perpetrate a temperature oscillation that leaves you alternately chilled and scalded.

The z transform of a delay of one sample is of course $1/z$. If the same delay is sampled twice as fast, the transfer function becomes $1/z^2$. Far from improving the digital control problem, an excessive sampling rate stacks up poles at the origin that seriously limit the possible speed of response; with a single such pole, dead-beat control is a possibility.

The analysis of a delay of two-and-a-half samples can be rather tricky, as you will have seen if you tried Exercise P11-3-1. On its own, the transform is zero—the sampler that builds the output sequence will miss any input impulses by half a sample width. With a single integrator, the transform is $1/z^2$ times $1/(z-1)$. The sampler cannot tell the difference between a delay of 2.1 samples and one of 2.9; samples 0, 1 and 2 are all zero, while from 3 onwards they are unity (see Fig. 11-23).

In general, the contribution of a delay T shorter than the sampling interval will depend heavily on the rest of the transfer function. It is best seen in the light of the state transition matrix. An input impulse will impart a step change to the state variables according to the appropriate column of B. By the time of the next sample, time $\tau - T$ later, this will have decayed to $\exp[A(\tau - T)]$ of its initial value. From then on each value is multiplied by $\exp(A\tau)$ and there just remains the task of summing the power series in $1/z$.

The choice of output DAC depends very much on the target performance. On/off control is adequate for an amazing variety of applications, particularly in domestic temperature regulation. However control of a robot or an autopilot is likely to warrant the extra cost of a DAC with a precision of at least eight bits. The accuracy of the DAC does not contribute greatly to the accuracy of the final output position; it is the accuracy of the feedback ADC which is important here. It does make a great difference to the 'ride' in terms of vibration and smoothness of control.

In the struggle for increased performance there is a temptation to overspecify the output device, be it servomotor or heater. It is always necessary first to analyse the possible failure modes, which will usually include a 'runaway condition'. An autopilot whose servomotor is powerful enough to pile on 30° of aileron in a fraction of a second is going to scare the hardiest of pilots, let alone the airline passengers. A fine compromise must be made between the maximum required manoeuvre and the limit of safe failure—the system should 'fail soft'.

If the computations within the controller are performed in floating-point arithmetic, loss of accuracy is only likely to arise if the computation is badly arranged to involve small differences of large quantities. If high speed and low cost are the objectives, however, the designer may be committed to the use of a dedicated chip with very limited arithmetic functions. There will be pressure to represent the variables as integers, and it might appear that one-byte precision would suffice.

Consider the transient response of the simple lag given by

X = X + (SIGNAL − X)/N

It is logical to make N a power of two, so that the multiplication can be

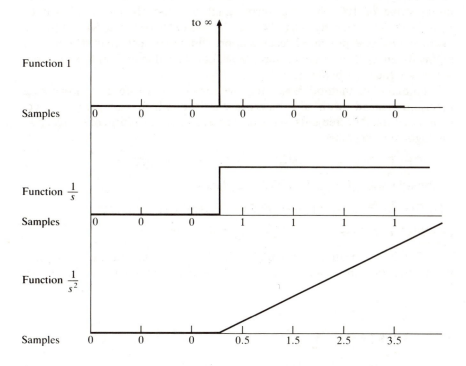

(a) For any non-multiple of τ, the z transform of a delayed inpulse is zero.

(b) For any delay between 2τ and 3τ, the z transform of a delayed step is

$$\frac{z^{-3}}{1 - z^{-1}} = \frac{1}{z^2(z - 1)}$$

(c) For delay $(2 + k)\tau$, $0 < k < 1$, the z transform
of a delayed ramp $\frac{1}{s^2} e^{-(2 + k)\tau s}$ is

$$z^{-3} \sum_0^\infty (n + 1 - k)z^{-n}$$

$$= z^{-3}\left[\frac{z}{(z - 1)^2} + (1 - k)\frac{z}{1 - z}\right]$$

$$= \frac{1}{z^2(z - 1)^2}\left[(1 - k)z + k\right]$$

Figure 11-23 A delay of 2.5τ changes the z transform in various ways.

replaced by an arithmetic shift—let us take $N = 16$. If X now starts with an initial value of 100, then for zero signal the first change in X will be by $-\text{INT}(100/16)$, i.e. by -6 to 94, then by -5 to 89, by -5 to 84 and by -5 again to 79. The exponential decay is approximated by a chain of straight-line segments. In itself this may be acceptable, but the value of X in this response will never then fall below 15.

To make the method reasonable, we really have to store X to two-byte precision instead, with a submerged 'iceberg' which can carry the accumulated small decrements to the part above water. In a form of *pseudocode*, the algorithm becomes:

(DHIGH, DLOW) ← (SIGNAL, 0) − (XHIGH, XLOW)

Arithmetic shift (DHIGH, DLOW) 4 places right

(XHIGH, XLOW) ← (XHIGH, XLOW) + (DHIGH, DLOW)

OUTPUT ← XHIGH

Now it is appealing to make $N = 256$ if possible, so that the 'shift right' merely becomes 'move high byte to low byte and copy sign to all bits of high byte'.

11-7 CONCLUSION

Practical controller design can involve as much art as science, particularly when there are several inputs and variables to be controlled. In an autopilot it was traditional to use the elevator to control height and the throttle to control airspeed. Suddenly it was realized that it is more logical to alter the glide angle to control the speed and to add extra thrust and energy to gain height; the roles of the throttle and aileron were reversed. Both control systems worked and maybe had the same poles, but their responses were subtly different.

Advanced analysis methods can aid the designer in the matter of choosing parameters, but they should not restrict the choice of alternative strategies out of hand. Even the selection of one damping factor over another may be as much a matter of taste as of defined performance. With the ability to add non-linearities at will, with the possibility of mixing discrete and continuous controllers which defy analysis, as a system designer you might be tempted to use a rule-of-thumb. Make sure that you first get to grips with the theory and gain familiarity through simulation and practice, so that when the time comes your thumb is well calibrated.

TWELVE

OPTIMAL CONTROL—NOTHING BUT THE BEST

12-1 INTRODUCTION

As soon as the controller develops beyond a simple feedback loop, the system designer is spoilt for choice. Where are the best positions to place the closed loop poles? How much integral control should be applied? These agonizing decisions can be resolved by looking for the 'best' choice of coefficients, those that will minimize a chosen *cost function*. This function might involve the integral of the square of the error at any time; if some multiple of the square of the drive signal is added in too, then the optimal solution can be a linear controller. But then the designer must agonize again over the particular choice of the cost function!

The cost function has a more tangible identity when we look at *single mission* tasks. An interceptor missile must close on an aircraft before its fuel expires and before the aircraft can move out of range. A lunar lander must bring the final altitude and vertical velocity to zero together, within close limits, while ensuring a safety reserve of fuel. A satellite must stabilize its attitude and station in orbit with a minimum of expenditure of reaction mass if it is to maximize its working life. In these examples, the choice of cost function is far from arbitrary. The controller which minimizes such a cost function is unlikely to rely on a simple linear feedback transfer function. We must develop a new range of methods.

12-2 DYNAMIC PROGRAMMING

Despite an impressive name, the principle of Bellman's Dynamic Programming is really quite straightforward. If the system has reached an intermediate point on an optimal path to some goal, then the remainder of the path must be the optimal path from that intermediate point to the goal.

It seems very simple, but its application can lead to some powerful theories. Consider the following illustration.

The car won't start, your train to the city leaves in twelve minutes' time, the station is nearly two miles away. Should you run along the road or take the perhaps muddy short cut across the fields? Should you set off at top speed, or save your wind for a sprint finish?

If you could see all possible futures, you could deduce the earliest possible time at which you might reach the station, given your initial state. Although there would be a best strategy that corresponded to this time, the best arrival time would be a function of state alone. Given your present fitness, distance from the station, muddiness of the field and of course time on the station clock, you can reach the station with one minute to spare. But every second you stand trying to decide on whether to take the short cut, one second of that minute ticks away.

You make the correct decision and start to sprint along the road. If an all-knowing computer clock could show your best arrival time, then the value displayed would stand still. You put on an extra burst of speed around the bend, but the computer display ticks forwards, not back; you are overtiring and will soon have to slow down. There is no way that you can pull that best arrival time backwards, because it is the optimum, the best that you can achieve in any circumstances. If you pace yourself perfectly, keeping exactly to the best route, you can hold the arrival time steady until you arrive.

You are over half-way there, hot, flustered, getting tired, five minutes later but in sight of the station. You have lost time. The computer clock stands at only thirty seconds to spare. That indicates the best arrival time from your new state. For each second that passes, you have to reduce the time required for the remainder of the journey by one second. If you could only see that display, you could judge whether a burst of effort was called for. If the computed time moves, the decision is wrong.

Now let us firm up the homespun philosophy into a semblance of mathematical reality. The cost is going to be the integral of a cost function, $c(\mathbf{x}, \mathbf{u})$. In this example the cost function is the constant, unity. Its integral gives the time taken for the journey. For your initial state, $\mathbf{x}$ at time t, there is a best possible cost $C(\mathbf{x}, t)$. You apply an input $\mathbf{u}$, taking you to a new state $\mathbf{x} + \delta\mathbf{x}$ after time δt. Now if you make the best possible time from your new state, your total cost will be

$$c(\mathbf{x}, \mathbf{u})\, \delta t + C(\mathbf{x} + \delta\mathbf{x}, t + \delta t)$$

You can influence this total cost by your choice of input $\mathbf{u}$. For optimal control you will select the value of $\mathbf{u}$ that minimizes the cost. However, the best you can do is to hold the value to be the same as that at the starting point; in other words

$$\min_{\mathbf{u}} \left[c(\mathbf{x}, \mathbf{u}) \, \delta t + C(\mathbf{x} + \boldsymbol{\delta}\mathbf{x}, t + \delta t) \right] = C(\mathbf{x}, t)$$

or $\quad \min_{\mathbf{u}} \left[c(\mathbf{x}, \mathbf{u}) \, \delta t + C(\mathbf{x} + \boldsymbol{\delta}\mathbf{x}, t + \delta t) - C(\mathbf{x}) \right] = 0$ (12-1)

For the best possible choice of input, the best remaining cost will reduce each instant by the same amount that the cost integral increases, i.e.

$$\min_{\mathbf{u}} \left[c(\mathbf{x}, \mathbf{u}) \, \delta t + \frac{dC}{dt} (\mathbf{x}, t) \, \delta t \right] = 0$$

Here the derivative of C is total, taking into account all the changes of the components of the state $\mathbf{x}$ in addition to any time dependence $\partial C/\partial t$.

We can take out a factor δt to give

$$\min_{\mathbf{u}} \left[c(\mathbf{x}, \mathbf{u}) + \frac{dC}{dt} (\mathbf{x}, t) \right] = 0$$ (12-2)

If we expand the total derivative in terms of the rates of change of all the states, this becomes

$$\min_{\mathbf{u}} \left[c(\mathbf{x}, \mathbf{u}) + \frac{\partial C}{\partial t} + \sum_i \dot{x}_i \frac{\partial C}{\partial x_i} \right] = 0$$ (12-3)

(If C is in fact time dependent, we can define an extra state variable with derivative unity, which will conveniently represent the time. The partial derivative $\partial C/\partial t$ can then be represented as the partial derivative with respect to a state variable, and can be lost inside the summation.)

We have two deductions to make. First, we must choose the input that will minimize this expression. Second, for this best input, the expression in the brackets will become zero.

Let us save a lot of typing effort by representing this expression to be minimized as G. To choose each component of the input, u_j, we must look at the partial derivative of G with respect to u_j. In some cases we might find a solution where this derivative can be zero; otherwise the input will be required to take a value at one or other of its limits.

With the cost function in the form of an integral, there is another way we can make the algebra neater. We can define a new state variable, x_0, which holds the value of the accumulated cost. The state equations of the system can be represented generally as

$$\dot{x}_i = f_i(\mathbf{x}, \mathbf{u})$$ (12-4)

for $i = 1$ to n, so now we have an extra equation

$$\dot{x}_0 = f_0(\mathbf{x}, \mathbf{u})$$

where f_0 is the same function as $c(\mathbf{x}, \mathbf{u})$.

Now we can substitute for the $\dot{x}$ terms using the state equations, to see that

$$G = \sum_{i=0}^{n} f_i(\mathbf{x}, \mathbf{u}) \frac{\partial C(\mathbf{x})}{\partial x_i}$$

The cost function has acquired an extra factor $\partial C/\partial x_0$, but this is obviously unity; an increase in the starting cost gives the same increase in the final cost. Taking the partial derivative of G with respect to any of the inputs we have

$$\frac{\partial G}{\partial u_j} = \sum_{i=0}^{n} \frac{\partial f_i}{\partial u_j} \frac{\partial C}{\partial x_i}$$

since C is not a function of $\mathbf{u}$. Either this will become zero in the allowable range of inputs, representing an acceptable minimum, or else the optimum input variable must take a value at one end or other of its limited range. In either event, when the optimal values of the inputs are substituted back into the expression for G to give the rate of change of total cost, the result must be zero.

Since after substitution G must be zero both throughout the trajectory and on any neighbouring optimal trajectory, its partial derivative with respect to a state variable will also be zero:

$$G = \sum f_i \frac{\partial C}{\partial x_i}$$

so

$$\frac{\partial G}{\partial x_j} = \sum_i \frac{\partial f_i}{\partial x_j} \frac{\partial C}{\partial x_i} + \sum_i f_i \frac{\partial^2 C}{\partial x_j \, \partial x_i} = 0 \qquad (12\text{-}5)$$

Now we have been weaving ever-increasing webs of algebra around the function C without really having a clue about what form it takes. It is an unknown, but not a complete mystery. We are starting to amass information not about the function itself but about its partial derivatives with respect to the states. Let us define these derivatives as variables in their own rights and see if we can solve for them. We define

$$p_i = \frac{\partial C}{\partial x_i}$$

so that we are trying to minimize $\Sigma p_i f_i$ with respect to the inputs. Since C is not a function of u, neither are the p's, so nothing is lost by regarding them as independent variables.

We might be able to solve for the p's by setting up differential equations for them. Try differentiating p_j:

$$\frac{dp_j}{dt} = \sum_i \dot{x}_i \frac{\partial p_j}{\partial x_i}$$

$$= \sum_i f_i \frac{\partial^2 C}{\partial x_i \, \partial x_j}$$

If C obeys certain continuity conditions, so that we may take liberties with the order of partial differentiation, Eq. (12-4) tells us that

$$\sum_i f_i \frac{\partial^2 C}{\partial x_i \, \partial x_j} = -\sum_i \frac{\partial f_i}{\partial x_j} \frac{\partial C}{\partial x_i}$$

so that

$$\frac{dp_j}{dt} = -\sum_i \frac{\partial f_i}{\partial x_j} \frac{\partial C}{\partial x_i}$$

$$= -\sum \frac{\partial f_i}{\partial x_j} p_i \tag{12-6}$$

Let us define a function H, which at first glance will look very much like our function G:

$$H = \sum_{i=0}^{n} f_i p_i \tag{12-7}$$

The subtle difference is that in H we regard the p's as independent variables, so that when we differentiate with respect to a state variable we have

$$\frac{\partial H}{\partial x_j} = \sum \frac{\partial f_i}{\partial x_j} p_i \tag{12-8}$$

while

$$\frac{\partial H}{\partial p_j} = f_j \tag{12-9}$$

Now we can express the state equations of (12-4) as

$$\dot{x}_i = \frac{\partial H}{\partial p_i}$$

while the p's are governed by the differential equations found by substituting definition (12-8) into Eq. (12-6):

$$\dot{p}_i = -\frac{\partial H}{\partial x_i}$$

We must of course minimize the *Hamiltonian* H by appropriate choice of inputs u, a task that will involve solving the *adjoint* differential equations for the variables p as well as solving for the state variables x. Not easy. The method is known (in a slightly different form) as *Pontryagin's maximum principle*. Let us see how it works in practice.

12-3 TIME-OPTIMAL CONTROL OF A SECOND-ORDER SYSTEM

Let us start gently, with the system $\ddot{y} = u$. The state equations are

$$\dot{x}_1 = x_2$$

$$\dot{x}_2 = u$$

The cost function is elapsed time, represented by x_0 where

$$\dot{x}_0 = 1$$

When we expand Eq. (12-7) and then substitute the functions that give the state derivatives we have

$$H = p_0 f_1 + p_2 f_2$$

$$= p_0 . 1 + p_1 x_2 + p_2 u$$

We must minimize H in terms of u, so u will take a value at one or other limit given by

$$u = u_{max} \, \text{sgn}(p_2)$$

All we have to do to apply time-optimal control is to solve for p_2. Now the differential equations for the p's are

$$\dot{p}_0 = -\frac{\partial H}{\partial x_0}$$

$$= 0$$

since H does not expressly involve x_0. In a similar way,

$$\dot{p}_1 = -\frac{\partial H}{\partial x_1} = 0$$

and

$$\dot{p}_2 = -\frac{\partial H}{\partial x_2} = -p_1$$

The first of these equations is satisfied by $p_0 = 1$, which is as it should be, while the second and third give

$$p_1 = a$$

$$p_2 = -at + b$$

So now we know that the sign of u is given by that of $at - b$. Have we solved the entire problem? These equations hold no more clues about the choice of the constants a and b. At most we can say that the optimal input involves applying full drive with at most one reversal of sign.

This form of analysis and that of *calculus of variations* are concerned with the manner of controlling the system, not with its details. A classic application of calculus of variations proves that the shortest distance between two points is a straight line. If I ask a villager how to get to the station by the shortest route, and I receive the reply 'Walk in a straight line', I am not altogether satisfied.

Nevertheless, the theory has narrowed down the search for an optimal strategy to just two decisions: the initial sign of the drive and the time at which it must reverse. If the target is zero error at zero velocity, then a strategy unfolds. The last part of the trajectory must bring the state to rest at the origin of the state space—more graphically seen as the origin of the phase plane, as described in Section 3-5. There will be two full-drive trajectories through the origin, one for each sense of drive. In this undamped case they will be parabolae. They divide the phase plane into two regions, as shown in Fig. 12-1, in which we must apply opposite signs of full drive.

Many methods have been devised for applying the appropriate drive signal, and in the days before it became easy to build a computer into the loop there was a benefit in ingenuity. The most obvious is the use of a non-linear function generator. A shaping circuit can compute the error multiplied by its modulus, which is then added to the velocity, applied to a comparator and the resulting sign passed to the drive. Alternatively, the position error signal can be added to a shaped velocity signal, representing the sign of the velocity

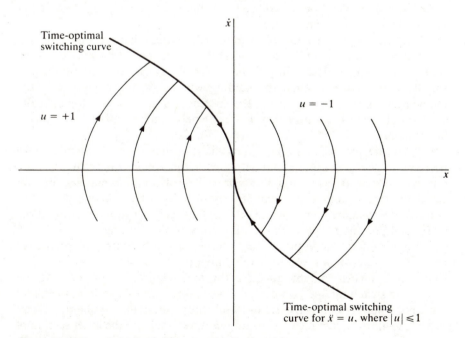

Figure 12-1 Time-optimal switching curve in the phase plane.

times the square root of its magnitude, and a comparator will give the same result.

A particularly interesting technique is termed *predictive control* and involves the use of a fast simulator model. Devised by John F. Coales and Maxwell Noton, it stimulated additional groups under Harold Chestnut and Boris Kogan to develop simplifications and enhancements. In effect, since there are few decisions to be made, a fast model can set out from the present state, be subjected to a trial input and be interrogated about the consequences. From the result, a new fast-model input is determined, and a decision is made about the input to the process.

Chestnut's second-order strategy is particularly simple. In the model, apply braking drive to bring the velocity to zero. If the model stops with a negative error then apply positive plant drive and vice versa.

For a three-integrator system, the drive may have two switches of sign. For n integrators it may have $n - 1$ switches. Predictive strategies to compute optimal control will rapidly become complicated, as interaction between the effects of the individual switching times is more complex and less predictable. One proposal involves models within models, each faster than the last.

Is optimal control really necessary or even desirable? It can be shown via the maximum principle that the fuel-optimal soft Lunar landing is achieved by burning the engines at full rate until touchdown. This involves igniting the engines at the last possible moment. Now as the spaceship approaches the moon before deceleration, it will be travelling in the region of one mile per second. If ignition is one second late. ...

The fuel usage is related closely to the momentum that must be removed, which is being continually augmented at the rate of one lunar gravity all the time the spaceship is descending. If the motors fire earlier and continue under slightly reduced thrust, then the landing will be slightly delayed and the momentum increased. However, the increase in use of fuel for a really substantial safety margin of thrust is extremely small—and well worth the investment.

There are predictive strategies for higher order systems which achieve performance close to optimal but with great reduction in complexity. At least one such system is flying in a Western satellite. Economic pundits running economic forecasting models can strictly be said to be applying predictive control. Their algorithms, however, are not altogether simple or straightforward, and you might question the optimality of their strategies!

For a system as simple as the humble position controller, predictive control is quite inappropriate. Even time-optimal control generated by a simple non-linearity might not give the most desirable response. At the origin, the switching line between positive and negative drive is vertical, and the input will buzz between extremes while the position dithers slightly. It can be beneficial to the lifetime of gears and motors to replace the inner section of a switching curve with a linear approximation of limited slope (see Fig. 12-2),

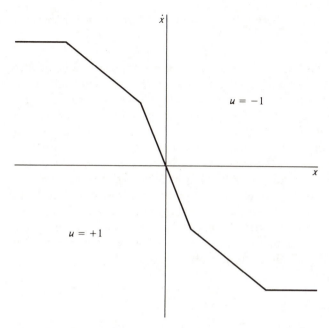

Figure 12-2 Piecewise linear suboptimal controller.

and even to reduce the gain of the comparator from infinity to a mere few thousands.

Even for a linear system with saturation, the linear system design guidelines are inappropriate. The slope of the position/velocity saturation line defines the time constant of a first-order type of response, an exponential decay with a time constant of, say, 0.1 seconds. For a tight loop gain, the linear region might have a second real root of one millisecond or less. We might have

$$s^2 + 1010s + 10\,000 = 0$$

which suggests a damping factor of five or more.

Any temptation to reduce the velocity term should be firmly resisted. The high loop gain arises not through a desire for faster response, but out of a need for precision. The constant term of 10 000 indicates that full drive will be applied if the position error exceeds 0.01 per cent of one unit.

Before leaving the maximum principle, let us look at the general problem of optimal control of a linear system. If

$$\dot{\mathbf{x}} = A\mathbf{x} + B\mathbf{u}$$

the Hamiltonian becomes

$$H = c(\mathbf{x}, \mathbf{u}) + \mathbf{p}'A\mathbf{x} + \mathbf{p}'B\mathbf{u} \qquad (12\text{-}10)$$

When we group the equations of (12-9) into matrix form, they become

$$\dot{\mathbf{p}} = -A'\mathbf{p}$$

where A' is the transpose of the state matrix A. For every eigenvalue of A in the stable left-hand half-plane, $-A'$ has one in the unstable half-plane. Solving the adjoint equations in forward time will be difficult, to say the least. Methods have been suggested in which the system equations are run in forward time, against a memorized adjoint trajectory, and the adjoint equations are then run in reverse time against a memorized state trajectory. The *boresight* method allows the twin trajectories to be *massaged* until they eventually satisfy the boundary conditions at each end of the problem.

When $c(\mathbf{x}, \mathbf{u})$ involves a term in u of second order or higher power, there can be a solution that does not require bang-bang control. The quadratic cost function is popular in that its optimization gives a linear controller. By going back to dynamic programming, we can find a solution without resorting to adjoint variables, although all is still not plain sailing.

12-4 QUADRATIC COST FUNCTIONS

Suppose we choose a cost function involving sums of squares of combinations of states, added to sums of squares of mixtures of inputs. We can exploit matrix algebra to express this mess more neatly as the sum of two *quadratic forms*:

$$\mathbf{x}'Q\mathbf{x} + \mathbf{u}'R\mathbf{u}$$

When multiplied out, each term above gives the required sums of squares and cross-products. The diagonal elements of R give multiples of squares of the u's, while the other elements define products of pairs of inputs. Without any loss of generality, Q and R can be chosen to be symmetric.

A more important property we must insist on, if we hope for proportional control, is that R is *positive definite.* The implication is that any non-zero combination of u's will give a positive value to the quadratic form. Its value will quadruple if the u's are doubled, and so the inputs are deterred from becoming excessively large. A consequence of this is that R is non-singular, so that it has an inverse.

For the choice of Q, we need only insist that it is positive semi-definite, that is to say no combination of x's can make it negative, although many combinations may make the quadratic form zero.

Having set the scene, we might start to search for a combination of inputs which would minimize the Hamiltonian of Eq. (12-10), now written as

$$H = \mathbf{x}'Q\mathbf{x} + \mathbf{u}'R\mathbf{u} + \mathbf{p}'A\mathbf{x} + \mathbf{p}'B\mathbf{u}$$

That would give us a solution in terms of the adjoint variables, p, which we would still be left to find. Instead let us look to the more fundamental relationship of (12-3) to see

$$\min_{\mathbf{u}} \left[c(\mathbf{x}, \mathbf{u}) + \frac{\partial C}{\partial t} + \sum_i \frac{\partial C}{\partial x_i} (\mathbf{x}, t)x_i \right] = 0$$

If the resulting control is going to be linear, then we are no longer in the dark about the nature of the 'best cost' function, C. If we start with all the initial state variables doubled, then throughout the resulting trajectory both the variables and the inputs will also be doubled. The cost clocked up by the quadratic cost function will therefore be quadrupled. We may, without much risk of being wrong, guess that the 'best cost' function must be of the form

$$C(\mathbf{x}, t) = \mathbf{x}'P\mathbf{x}$$

Now if the endpoint of the integration is in the infinite future, it does not matter when we start the experiment, so we can assume that the matrix P is a constant. If there is some fixed end time, however, so that the time of starting affects the best total cost, then P will be a function of time, $P(t)$.

To deal with a set of partial derivatives one at a time can be cumbersome, but an algebraic labour-saving device is available. The *operator* ∇ denotes partial differentiation with respect to a complete vector of variables, giving a result that is the vector made up of the individual partial derivatives. The definition means that

$$\nabla_x C = \left(\frac{\partial C}{\partial x_1}, \frac{\partial C}{\partial x_2}, \ldots \right)$$

It is not difficult to show that

$$\nabla_x(\mathbf{x}'P\mathbf{x}) = \mathbf{x}'(P + P')$$

which if P is symmetric equals

$$2\mathbf{x}'P$$

We can write the expression to be minimized as

$$c + \frac{dC}{dt} = c + \frac{\partial C}{\partial t} + \nabla_x C \dot{\mathbf{x}}$$

$$= c(\mathbf{x}, \mathbf{u}) + \mathbf{x}'\dot{P}\mathbf{x} + 2\mathbf{x}'P\dot{\mathbf{x}}$$

$$= c(\mathbf{x}, \mathbf{u}) + \mathbf{x}'\dot{P}\mathbf{x} + 2\mathbf{x}'P(A\mathbf{x} + B\mathbf{u})$$

$$= \mathbf{x}'Q\mathbf{x} + \mathbf{u}'R\mathbf{u} + \mathbf{x}'\dot{P}\mathbf{x} + 2\mathbf{x}'P(A\mathbf{x} + B\mathbf{u}) \quad (12\text{-}11)$$

To look for a minimum of this with respect to the inputs, we must differentiate with respect to each u and equate the expression to zero. It is another case for the use of the ∇ operator:

$$\nabla_u \left(\frac{dC}{dt} \right) = 2\mathbf{u}'R + 2\mathbf{x}'PB$$

When we equate this to zero, we have

$$2u'R + 2x'PB = 0$$

and so

$$\mathbf{u} = -R^{-1}B'P\mathbf{x}$$

It is a clear example of proportional feedback, but we must still put a value to the matrix P. When we substitute for u back into Equation (12-11) we must get the answer zero, so

$$x'Qx + x'PBR^{-1}RR^{-1}B'Px + x'\dot{P}x + 2x'PAx - 2x'PBR^{-1}B'Px = 0$$

that is

$$\mathbf{x'}(Q + PBR^{-1}B'P + \dot{P} + 2PA - 2PBR^{-1}B'P)\mathbf{x} = 0$$

This must be true for all states, x, and so we can equate the resulting quadratic to zero term by term. It is less effort to make sure that the matrix in the brackets is symmetric, and then to equate the whole matrix to the zero matrix. If we split $2PA$ into the symmetric form $PA + A'P$ (equivalent for quadratic form purposes), we have

$$\dot{P} + PA + A'P + Q - PBR^{-1}B'P = 0$$

This is the matrix *Riccati* equation, and much effort has been spent in its systematic solution. In the infinite-time case, where P is constant, the quadratic equation in its elements can be solved with a little labour.

Is this effort all worth while? We can apply proportional feedback, where with only a little effort we choose the locations of the closed loop poles. These locations may be arbitrary, so we seek some justification for their choice. Now we can choose a quadratic cost function and deduce the feedback which will minimize it. This cost function may itself be arbitrary, and its selection will almost certainly be influenced by whether it will give 'reasonable' closed loop poles!

Exercise 12-4-1 *Find the feedback that will minimize the integral of $y^2 + a^2u^2$ in the system $y = u$.*

Exercise 12-4-2 *Find the feedback that will minimize the integral of $y^2 + b^2y^2 + a^2u^2$ in the system $y = u$.*

Before reading the solutions which follow, try the exercises yourself. The first problem is extremely simple, but demonstrates the working of the theory. In the matrix state equations and quadratic cost functions, the matrices reduce to 1 by 1 size, where

$$A = 0$$
$$B = 1$$
$$Q = 1$$
$$R = a^2, \text{ so } R^{-1} = 1/a^2$$

There is no time limit specified; therefore $dP/dt = 0$. We then have the equation

$$PA + A'P + Q - PBR^{-1}B'P = 0$$

to solve for the 'matrix' P, here 1 by 1. Substituting, we have

$$0 + 0 + 1 - p . 1\left(\frac{1}{a^2}\right) . 1p = 0$$

that is

$$p^2 = a^2$$

Now the input is given by

$$u = -R^{-1}B'Py$$
$$= -\left(\frac{1}{a^2}\right) . 1ay$$
$$= -\frac{y}{a}$$

and we see the relationship between the cost function and the resulting linear feedback.

The second example is a little less trivial, involving a second-order case. We now have 2 by 2 matrices to deal with, and taking symmetry into account we are likely to end up with three simultaneous equations as we equate the components of a certain matrix to zero. Now if we take y and $\dot{y}$ as state variables we have

$$A = \begin{bmatrix} 0 & 1 \\ 0 & 0 \end{bmatrix}$$

$$B = \begin{bmatrix} 0 \\ 1 \end{bmatrix}$$

$$Q = \begin{bmatrix} 1 & 0 \\ 0 & b \end{bmatrix}$$

and

$$R = a^2$$

The matrix P will be symmetric, so we can write

$$P = \begin{bmatrix} p & q \\ q & r \end{bmatrix}$$

Once again dP/dt will be zero, so we must solve

$$PA + A'P + Q - PBR^{-1}B'P = 0$$

Thus

$$\begin{bmatrix} 0 & p \\ 0 & q \end{bmatrix} + \begin{bmatrix} 0 & 0 \\ p & q \end{bmatrix} + \begin{bmatrix} 1 & 0 \\ 0 & b \end{bmatrix} - \begin{bmatrix} p & q \\ p & r \end{bmatrix}\begin{bmatrix} 0 \\ 1 \end{bmatrix}\frac{1}{a^2} \begin{bmatrix} 0 & 1 \end{bmatrix}\begin{bmatrix} p & q \\ q & r \end{bmatrix}$$

$$= \begin{bmatrix} 0 & 0 \\ 0 & 0 \end{bmatrix}$$

that is

$$\begin{bmatrix} 1 & p \\ p & b + 2q \end{bmatrix} - \frac{1}{a^2}\begin{bmatrix} q^2 & qr \\ qr & r^2 \end{bmatrix} = \begin{bmatrix} 0 & 0 \\ 0 & 0 \end{bmatrix}$$

from which we deduce that

$$q^2 = a^2$$

$$qr = a^2 p$$

and

$$r^2 = a^2(b + 2q)$$

so that $q = a$ (the positive root applies), $r = a\sqrt{2a + b}$ and $p = \sqrt{2a + b}$.
Now u is given by

$$u = -R^{-1}B'P\mathbf{x}$$

$$= -1/a^2 \begin{bmatrix} 0 & 1 \end{bmatrix}\begin{bmatrix} \sqrt{2a + b} & a \\ a & a\sqrt{2a + b} \end{bmatrix}\begin{bmatrix} x_1 \\ x_2 \end{bmatrix}$$

$$= -\frac{1}{a}x_1 - \frac{\sqrt{2a + b}}{a}x_2$$

It seems a lot of work to obtain a simple result. There is one very
interesting conclusion, however. Suppose that we are concerned only with the
position error and do not mind large velocities, so that the term b in the cost
function is zero. Now our cost function is simply given by the integral of the
square of error plus a multiple of the square of the drive. The equation for the
drive becomes

$$u = -\frac{1}{a}x_1 - \frac{\sqrt{2}}{\sqrt{a}}x_2$$

$$= -\frac{1}{a}y - \frac{\sqrt{2}}{\sqrt{a}}\dot{y}$$

so that the closed loop behaviour is described by

$$\ddot{y} = u$$

$$= -\frac{1}{a} y - \frac{\sqrt{2}}{\sqrt{a}} \dot{y}$$

that is

$$\ddot{y} + \frac{\sqrt{2}}{\sqrt{a}} \dot{y} + \frac{1}{a} y = 0$$

Perhaps there is a tangible reason for placing closed loop poles to give a damping factor of 0.707 after all.

12-5 IN CONCLUSION

Control theory exists as a fundamental necessity if we are to devise ways of persuading dynamic entities to do what we want them to. By searching for state variables, we can set up equations with which to simulate the system's behaviour with and without control. By applying a battery of mathematical tools we can devise controllers that will meet a variety of objectives, and some of them will actually work. Others will spring from high mathematical ideals, seeking to extract every last ounce of performance from the system, and might neglect the fact that a motor cannot reach infinite speed or that a computer cannot give an instant result.

Care should be taken before putting a control scheme into practice. Once the strategy has been fossilized into hardware, changes can become expensive. You should be particularly wary of accepting that a simulation success is evidence that a strategy will work, especially when both strategy and simulation are digital:

A digital simulation of a digital controller will perform exactly as you expect it will—however catastrophic the control may be when applied to the real world.

You should by now have a sense of familiarity with many aspects of control theory, especially in the foundations in time and frequency domain and in methods of designing and analysing linear systems and controllers. Many other topics have not been touched on here: system identification, optimization of stochastic systems and model reference controllers are just a start. The subject is capable of enormous variety, while a single technique can appear in a host of different mathematical guises.

To become proficient at control system design, nothing can improve on practice. Algebraic exercises are not enough; your experimental controllers should be realized in hardware if possible. Examine the time responses, the stiffness to external disturbance, the robustness to changing parameter values. Then read more of the wide variety of books on general theory and special topics.

NOW READ ON

A solid introduction to both frequency domain and state space methods can be found in *Automatic Control Systems* by Benjamin C. Kuo, published by Prentice-Hall, 5th edn, 1987, ISBN 0-13-055070-1.

If you take kindly to the exercise-drill approach, you might like to look at *Schaum's Outline of Feedback and Control Systems*, by J. J. DiStefano *et al.*, McGraw-Hill, 1967, ISBN 0-07-017045-2.

Francis H. Raven puts some emphasis on the use of the computer for understanding control fundamentals in *Automatic Control Engineering*, McGraw-Hill, 4th edn, 1987, ISBN 0-07-051233-7.

John D'Azzo and Constantine H. Houpis have added some optimal control to the 3rd edition of their classic *Linear Control System Analysis and Design*, McGraw-Hill, 3rd edn, 1988, ISBN 0-07-016186-0.

For discrete-time systems, C. H. Houpis has teamed up with G. B. Lamont to produce *Digital Control Systems: Theory, Hardware, Software*, McGraw-Hill, 1985, ISBN 0-07-030480-7. You will have to buy the solutions manual separately as ISBN 0-07-030481-5.

Paul Katz presents a terse study of discrete-time methods in *Digital Control using Microprocessors*, Prentice-Hall, 1981, ISBN 0-13-212191-3. It makes a good reference book.

An early classic on optimal control is by M. Athans and P. L. Falb, *Optimal Control*, McGraw-Hill, 1966, ISBN 0-07-002413-8.

Many excellent books on control have not been mentioned here. Find a reputable bookshop or library and browse until you find a book in a style which suits your taste.

APPENDIX A

PLOTTING THE ROOT LOCUS

A-1 INTRODUCTION

Figure A-1 shows the listing of a computer program written in **GWBASIC** which will plot a root locus on the screen. Using the **GRAPHICS** command beforehand will configure the Print Screen button to cause the screen image to be sent to an Epson dot-matrix printer; that is how most of the root locus diagrams in this book have been prepared.

This appendix breaks the listing down into its components, partly to make it easier to understand and more importantly so that you can modify it for your own particular purposes. You can also make any changes necessary for another dialect of BASIC or if your machine has a different graphics configuration.

A-2 PREPARATION

Line 5 sets up the screen mode with SCREEN 3. It also defines the range of coordinates as 640 horizontal and 400 vertical. If your machine does not support this mode, then edit the line appropriately. The variable ASPECT is a 'fiddle factor' so that the unit circle will appear truly circular on the screen-dump to the printer.

Line 10 determines the range of the complex frequency over which the

```
5   SCREEN 3:CLS:XRANGE=640:YRANGE=400:ASPECT=.825
10  SMIN=-1:SMAX=1:N=64:SS=(SMAX-SMIN)/N:SLIM=SS*SS/4
15  SSC=XRANGE/(SMAX-SMIN):X0=-SMIN*SSC
20  WSC=ASPECT*SSC:Y0=YRANGE*3/4:PI=3.14159
25  WMAX=Y0/WSC:WMIN=-WMAX/3:WW=(WMAX-WMIN)/N
30  INPUT"how many poles ";NP:DIM P(NP,1)
40  FOR I=1 TO NP:INPUT"real, imag ";P(I,0),P(I,1):NEXT
50  INPUT"how many zeroes ";NZ:DIM Z(NZ,1)
60  FOR I=1 TO NZ:INPUT"real, imag ";Z(I,0),Z(I,1):NEXT
65  INPUT"Unit circle (y/n) ";C$
70  G0=1:CLS
80  LINE(0,Y0)-(XRANGE,Y0),,,&H1111:REM Real axis
85  LINE(X0,0)-(X0,YRANGE),,,&H1111:REM Imaginary axis
90  IF C$<>"Y" AND C$<>"Y" THEN LINE(X0-SSC,Y0)-(X0-SSC,Y0+10):GOTO 100
95  FOR I=0 TO 2*PI STEP .04:PSET(X0+SSC*COS(I),Y0-WSC*SIN(I)):NEXT
100 W=WMIN+WW/2:S=SMIN:DS=0:DW=WW
110 FOR X=0 TO N:FOR Y=0 TO N:GOSUB 1000:REM Gain GR +jGI
115 REM:IF GI<0 THEN PSET(X0+S*SSC,Y0-W*WSC)
120 IF Y>0 AND GI*GIOLD<=0 AND GR<0 THEN GOSUB 2000:REM Plot locus segment
140 GROLD=GR:GIOLD=GI:W=W+WW:NEXT
150 S=S+SS:W=WMIN+WW/2:NEXT
200 W=WMIN:S=SMIN:DS=SS:DW=0
210 FOR Y=0 TO N:FOR X=0 TO N:GOSUB 1000:REM Gain GR +jGI
220 IF X>0 AND GI*GIOLD<=0 AND GR<0 THEN GOSUB 2000:REM Plot locus segment
240 GROLD=GR:GIOLD=GI:S=S+SS:NEXT
250 W=W+WW:S=SMIN:NEXT
295 REM Mark poles with small circles, zeros with larger.
300 FOR I=1 TO NP:CIRCLE(X0+P(I,0)*SSC,Y0-P(I,1)*WSC),3:NEXT
310 FOR I=1 TO NZ:CIRCLE(X0+Z(I,0)*SSC,Y0-Z(I,1)*WSC),5:NEXT
320 END
1000 NR=G0:NI=0:DR=1:DI=0:REM Calculate gain GR+j*GI at freq S+j*W
1010 FOR I=1 TO NP:VR=S-P(I,0):VI=W-P(I,1):DR1=DR*VR-DI*VI
1020 DI=DR*VI+DI*VR:DR=DR1:NEXT:MD=DR*DR+DI*DI
1030 IF NZ=0 THEN 1060
1040 FOR I=1 TO NZ:VR=S-Z(I,0):VI=W-Z(I,1):NR1=NR*VR-NI*VI
1050 NI=NR*VI+NI*VR:NR=NR1:NEXT
1060 GR=(NR*DR+NI*DI)/MD:GI=(NI*DR-NR*DI)/MD
1070 RETURN
2000 DGI=GI-GIOLD:REM Interpolate to find crossing of GI=0
2010 IF DGI=0 THEN R=.5 ELSE R=ABS(GI/DGI)
2020 CS=S-DS*R:CW=W-DW*R:REM estimated crossing frequency
2030 DR=0:DI=0:REM These will hold sum of 1/(s-zeros)-1/(s-poles)
2040 FOR I=1 TO NP:VR=CS-P(I,0):VI=CW-P(I,1):VM=VR*VR+VI*VI
2050 IF VM>SLIM THEN DR=DR-VR/VM:DI=DI+VI/VM:ELSE RETURN
2060 NEXT
2080 FOR I=1 TO NZ:VR=CS-Z(I,0):VI=CW-Z(I,1):VM=VR*VR+VI*VI
2090 IF VM>SLIM THEN DR=DR+VR/VM:DI=DI-VI/VM:ELSE RETURN
2100 NEXT
2110 DM=SQR(DR*DR+DI*DI):IF DM<SSC/XRANGE/N    THEN RETURN
2120 LS=SS*DR/DM/2:LW=-SS*DI/DM/2
2130 LINE(X0+SSC*(CS-LS),Y0-WSC*(CW-LW))-(X0+SSC*(CS+LS),Y0-WSC*(CW+LW))
2140 RETURN
```

Figure A/1 Listing of BASIC root locus program using gradient computation.

locus will be plotted. N gives the resolution in evaluated points across the screen, while SLIM is set up for later use. In lines 15 to 25 the imaginary range is set up to correspond to the real, the positive range being three times the negative.

Lines 30 to 60 request the user for the pole and zero coordinates. The axes are then plotted as broken lines in 80 and 85 and the unit circle is added if requested. Now we are ready for the plot.

A-3 PLOTTING THE LOCUS

In Sec. 7-2 we saw that a 'character plot' could display the root locus as the boundary between a region where the imaginary part of the gain function was positive and one where it was negative. Leave out the characters and display just the boundary and we have a root locus.

At the simplest level, that is all there is to it. The complex gain function is evaluated at each point of a grid in the s plane. If the sign of the imaginary part changes between one point of the grid and the next, then a point is plotted. Lines 100 and 110 set up a grid in which vertical sweeps of W, the imaginary variable, are made for increasing steps of s, the real. Subroutine 1000 evaluates the complex gain for each complex frequency, returning the gain as GR + j GI, and GI is compared against its previous value stored in GIOLD. If GI has changed sign and if GR is negative, and if this is not the first point of the sweep, then a mark is made using subroutine 2000.

In lines 200 to 250 the grid is covered again, this time with horizontal sweeps. In this way a vertical feature cannot be missed and is sure to be traced out. Finally the poles and zeros are marked on the plot.

A-4 THE SUBROUTINES

The complex gain is evaluated exactly as described in the text, by taking the product of the complex vectors from the zeros to the complex frequency, by computing also the product of the vectors from the poles to the complex frequency and then by dividing the first of these by the second. This is all achieved in lines 1000 to 1070.

The subroutine at 2000 is a little more involved. It would have been possible to plot a point midway between the new complex frequency and the previous point, but we can get a more accurate result by interpolating between them according to the values found for their imaginary gain components. CS and CW are found by subtracting an appropriate proportion of the complex step length (real or imaginary according to the direction of sweep) from the latest grid point. Once again this point could simply be plotted as a spot, but we can do even better.

The root locus is the locus of complex frequency points which map into the negative real gain axis. A short line segment in the frequency plane will map into a segment in the gain plane, its length magnified and its direction rotated according to the derivative of the gain function. If, conversely, we divide a small segment of the negative real gain axis by the complex number representing this derivative, then we will have a line segment in the frequency plane giving the direction of the root locus.

Now to a first approximation we can use the ratio of the change in complex gain divided by the change in frequency between this grid point and the last. The result is acceptable if the grid points are close together, but this can be slow to plot. Near poles, zeros and branch points, the plot tends to 'fray at the edges'.

Lines 2030 onwards calculate the argument of the derivative, the angle through which the line segment is turned. The gain is the product of numerator terms $s - z_r$. Differentiate the product 'by parts', and each term gives a contribution which is the rest of the expression for the gain with that term omitted. The contribution is the gain function divided by that term.

Similarly the denominator is the product of terms $s - p_r$. Again the derivative by parts gives a contribution for each term, being the gain divided again by that term, but this time the sign of the contribution is negated. The derivative can be calculated as

$$G(s) \left[\sum_r \frac{1}{s - z_r} - \sum_r \frac{1}{s - p_r} \right]$$

In lines 2040 to 2100 these sums are evaluated, needing only to be multiplied by the gain $G(s)$ to give the derivative. However, if we have interpolated correctly, the value of $G(s)$ at (CS, CW) should be real. We therefore rotate DS + j DW by minus the argument of the sum of the inverted vectors to get a line segment to plot.

So why did we set up SLIM? If the grid point is too close to a pole or zero we can have trouble. If we are closer than SLIM then the line segment is not plotted. Similarly, if the gradient is too close to zero, as at a branch point, then it is safer to leave a gap in the locus than to plot a bristle of disjointed segments.

APPENDIX B

SOLUTIONS TO EXERCISES

Many of the exercises are solved in the section that follows them. Some other solutions appear in later chapters. It is worth giving the problems a good try before giving in.

Exercise 3-5-1 The solution is shown in Fig. B/3-5-1.

Exercise 3-6-1 The solution follows in the text.

Exercise P5-3-1 The solution follows in the text.

Exercise P5-4-1 Put $u = \text{real}(-j\, e^{(-1+2j)t})$ to find

$$x = \tfrac{1}{8} e^{-t}[\cos(2t) - \sin(2t)]$$

Exercise P5-4-2 The input corresponds to a pole of the system. The input must include $t \exp[(-2+j)t]$ (see Sec. 5-9).

Exercise 5-7-1 The solution follows in the text.

Exercise P6-2-1 The solution follows in the text.

Exercise P6-3-1 $\dfrac{\partial u}{\partial x} = 2x = \dfrac{\partial v}{\partial y}$ $\qquad \dfrac{\partial v}{\partial x} = 2y = -\dfrac{\partial u}{\partial y}$

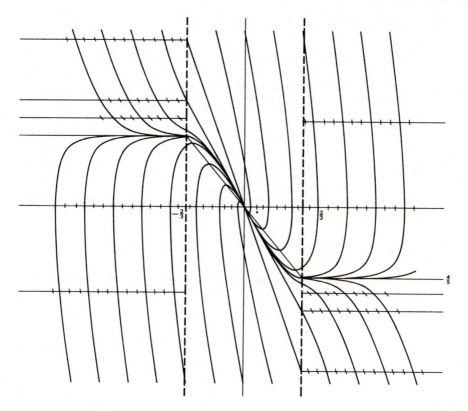

In the saturation regions $\ddot{x} = -5\dot{x} \pm 4$

$$\frac{\mathrm{d}\dot{x}}{\mathrm{d}x} = -5x \pm \frac{4}{\dot{x}}$$

Figure B/3-5-1 Phase plane solution of Exercise 3-5-1.

Exercise P6-5-1 $x(t) = [A + x(0)] \sin(t) + [B + x(0)] \cos(t) + C\,e^{-at}$
where $A = 5a/(1 + a^2)$, $B = -5/(1 + a^2)$ and $C = 5/(1 + a^2)$.

Exercise P6-6-1
$$F(\mathrm{j}\omega) = \int_{-\infty}^{\infty} f(t)\,e^{-\mathrm{j}\omega t}\,\mathrm{d}t$$

$$= \int_{0}^{\infty} e^{-at}\,e^{-\mathrm{j}\omega t}\,\mathrm{d}t$$

[since $f(t) = 0$ for $t < 0$]

$$= \frac{1}{a + \mathrm{j}\omega}$$

Exercise 6-2-1 The solution follows in the text.

Exercise 6-2-2 The 'curly squares' plot is shown in Fig. B/6-2-2.

Exercise 6-6-1 The phase plot is shown in Fig. B/6-6-1(a). The maximum feedback gain for stability is 4.25. The transfer function is

$$\frac{1}{s(s + 0.25)(s + 4)}$$

The Nichols plot is shown in Fig. B/6-6-1(b).

Exercise 6-6-2 The complete diagram shows that the -1 point is encircled for all gains (see Fig. B/6-6-2).

Exercise 7-4-1 See Fig. B/7-4-1.

Exercise 7-4-2 The asymptotes are along the positive and negative imaginary axes (see Fig. B/7-4-2).

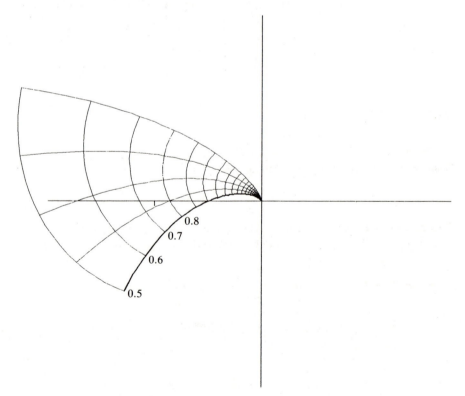

Figure B/6-2-2 Extended 'curly squares' Nyquist diagram for $1/s(s + 1)^2$.

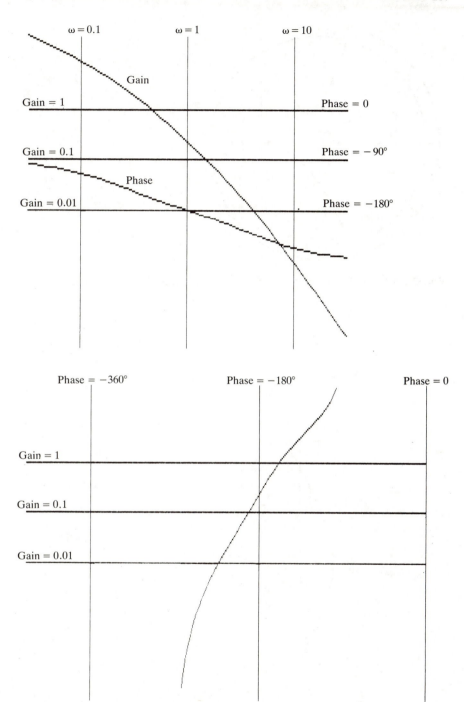

Figure B/6-6-1 (*a*) Bode plot of gain and phase. (*b*) Nichols plot.

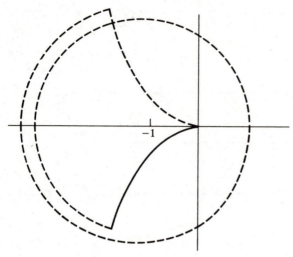

Figure B/6-6-2 Whiteley plot of $1/s^2(s + 1)$. The -1 point is encircled for all gains.

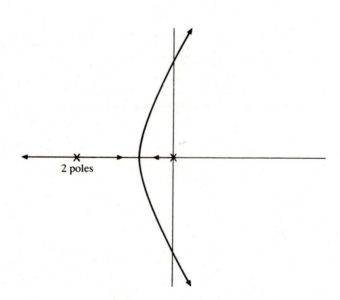

Figure B/7-4-1 Root locus of $1/s(s + 1)^2$.

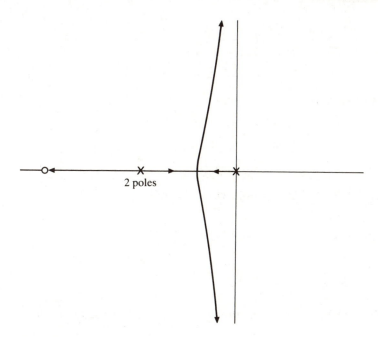

Figure B/7-4-2 Root locus of $(s + 2)/s(s + 1)^2$.

Exercise 7-6-1 The roots lie on the imaginary axes—steady oscillation.

Exercise 7-6-2 The solution follows in the text.

Exercise 7-6-3 The solution follows in the text.

Exercise P8-3-1 The solution follows in the text.

Exercise P8-4-1 The product of the roots is in fact $\det(A) (-1)^n$.

Exercise 8-3-1 $Y = 6/(s^2 + 5s + 6)U$.

Exercise 8-3-2 $Y = \begin{bmatrix} \dfrac{1}{s + 3} & \dfrac{1}{s + 2} \end{bmatrix} \begin{bmatrix} U_1 \\ U_2 \end{bmatrix}.$

Exercise 8-6-1
$$y(s) = \frac{s^2}{s^2 + 2s + 1} U(s)$$

$$= \left(1 - \frac{2s + 1}{s^2 + 2s + 1} \right) U(s)$$

If we construct x_1 to satisfy

$$\ddot{x}_1 + 2\dot{x}_1 + x_1 = u$$

then

$$y = u - 2\dot{x}_1 - x_1.$$

In companion form we have

$$\dot{x}_1 = x_2$$

$$\dot{x}_2 = -x_1 - 2x_2 + u$$

$$y = -x_1 - 2x_2 + u$$

This can be approximated by software which includes a loop with

```
DX1 = X2
DX2 = -X1 - 2*X2 + V
Y = DX2
X1 = X1 + DX1*DT
X2 = X2 + DX2*DT
```

Exercise P9-3-1 By more formal methods you will reach the same result of

$$\begin{bmatrix} -2 & 0 \\ 0 & -3 \end{bmatrix}$$

found in Sec. 3-4.

Exercise 9-2-1

$$\frac{1}{(s+2)(s+3)} = \frac{1}{s+2} - \frac{1}{s+3}$$

$$y = z_1 - z_2$$

where

$$\dot{z}_1 = -2z_1 + u$$

$$\dot{z}_2 = -3z_1 + u$$

Exercise 9-2-2 See Sec. 9-3.

Exercise 9-3-1 Eigenvalues are $\pm j$.
Jordan form is

$$\begin{bmatrix} \dot{x}_1 \\ \dot{x}_2 \end{bmatrix} = \begin{bmatrix} j & 0 \\ 0 & -j \end{bmatrix} \begin{bmatrix} x_1 \\ x_2 \end{bmatrix} + \begin{bmatrix} 1 \\ 1 \end{bmatrix} u$$

$$y = \frac{1}{2_j}(x_1 - x_2)$$

Alternative real form is

$$\begin{bmatrix} \dot{x}_1 \\ \dot{x}_2 \end{bmatrix} = \begin{bmatrix} 0 & 1 \\ -1 & 0 \end{bmatrix} \begin{bmatrix} x_1 \\ x_2 \end{bmatrix} + \begin{bmatrix} 0 \\ 1 \end{bmatrix} u$$

$$y = x_1$$

Exercise 9-4-1 Controllable.

Exercise 9-4-2 Not controllable—of rank 1.

Exercise 9-4-3 Controllable.

Exercise 9-4-4 Worked in the text which follows (see also Fig. 9-7).

Exercise 9-5-1 The solution follows in the text.

Exercise 9-5-2 $p = q = 4$, $u = -100y - 20\hat{y}$ (see Fig. B/9-5-2).

Exercise 9-6-1 The solution follows in the text.

Exercise 9-6-2 The solution follows in the text.

Exercise 10-3-1 Far from it! The eigenvalues are -1 and -3.

Exercise 10-3-2 The transform of the product is not equal to the product of the transforms (see Sec. P11-3 for an explanation).

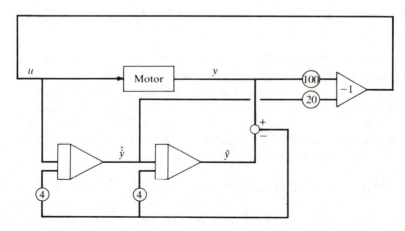

Figure B/9-5-2 Schematic solution to Exercise 9-5-2.

Exercise 10-3-1   See Sec. 10-4.

Exercise 10-4-1   The solution follows in the text.

Exercise 10-6-1   The solution follows in the text.

Exercise 10-6-2   The solution follows in the text.

Exercise 10-7-1 $1 + \tau + \tau^2/2 + \tau^3/6 + \tau^4/24$.

Exercise 11-3-1   (a) $1/(z - 1)$ is more realistic than $z/(z - 1)$, since it has no response until after $t = 0$. (b) $(1 - e^{-a\tau})z/a(z - e^{-a\tau})(z - 1)$. (c), (d) and (e) see Fig. 11-23.

Exercise 11-5-1   See Sec. 11-4.

Exercise 11-2-1   The solution follows in Sec. 11-3.

Exercise 11-2-2   The solution follows in Sec. 11-4.

Exercise 11-3-1   The solution follows in the text.

Exercise 11-4-1   The solution follows in the text.

Exercise 11-5-1   The response to a step of disturbance has final value given by the limit of $1/[1 + G(s)H(s)]$ as s tends to zero. In the first case this is $s(s + 1)/(s^2 + s + 1)$, which tends to zero. In the second case it is $(s + 1)/(s + 2)$, which tends to $\frac{1}{2}$.

Exercise 11-5-2   (See the exercise above.) By setting $H(s) = (s + a)/s$ we obtain $s(s + 1)/(s^2 + 2s + a)$, which tends to zero.

Exercise 11-5-3   In principle, control using a count of 'error stripes' will be hopeless, since steady settling must be obtained to within a very small fraction of a stripe and no adequate digital velocity signal can be derived for an error of less than one stripe. In practice, it is possible to add a central linear analogue control zone by using the 'raw' photocell signal. For small deflections this is proportional to the position error, and phase advance can be used to stabilize the response.

Exercise 12-4-1   The solution follows in the text.

Exercise 12-4-2   The solution follows in the text.

INDEX